Skip the Typing Test – I'll Manage the Software:

One Woman's Pioneering Journey in High Tech

Beverly Schultz

Hardcover ISBN: 978-1-62646-693-7
Paperback ISBN: 978-1-62646-694-4

Published by BookLocker.com, Inc., Bradenton, Florida.

Printed in the United States of America.

BookLocker.com, Inc.
2013

First Edition

This book is dedicated to:

my husband, Ken, who encouraged

and supported a developmental engineer so she could thrive

and my children Eric, Melissa and Michael,

who blossomed because of and in spite of

my being a working mother

Acknowledgments

People described in this book are real people as I remember them. Working with such strong people in my career has made them essential to my story, and I value their contribution to my life. Their names have been changed to protect their privacy.

Whatever name is attached to them, I thank them for being my partners, my mentors, and my teams as we worked together to accomplish great things.

Companies described in this book are near and dear to me. I gave a good part of my life to each of them, and I shared my soul with them. Had I worked for different companies, the story would have been a little different, but the theme of how I was stretched and molded by my work doesn't depend on a company, but on the workings that could be attributed to companies across the board in this time period. What I've said about these companies is my personal opinion only, and may not be in any way be true.

Special thanks to my editor, Dorie McKeeman, who held my hand through many rounds of optimizing. Her insight and her persistence were invaluable.

Prologue

It was 1966. My first year of teaching mathematics was finished, and it was a great year! I had been nominated for "Outstanding Teacher of the Year"—a wonderful feeling. My husband, Ken, was transferred to Colorado Springs, Colorado, so I suddenly found myself job hunting. It was June, and fall slots for teachers were pretty much filled. I saw an ad in the paper and called to set up an interview with Kamen Nuclear, as they seemed like an interesting company.

What an *exciting* day of interviewing! I spoke only to men, and each one grilled me relentlessly. One made me write math equations of things he was talking about. One had me correct the equations on a problem he was working on. They proposed challenging problems and I had to solve them. They told me about their work. I felt wrung out, but by the end of the day I *loved* this company and what they were doing. When they brought me back to the head of the group, he said, "The men here loved you and think you'll fit right in. We'd like to offer you a job with us!" I was elated. Then he continued, "There is some paperwork for you to sign, and as a formality, of course, you'll have to take the typing test."

They wanted to hire me not as their colleague, but as their secretary, because I was a woman. My degree in mathematics was nice; that meant I could correct the equations and fix up the papers of these guys while they got to do the fun stuff! *I* wanted to do the fun stuff! I didn't want to be a secretary! I refused to take the typing test and left, deflated.

That week I got another job offer—to teach—and so I forgot about Kamen Nuclear and remained a teacher for the next few years. But I didn't forget that there were exciting jobs to be had that I would like to try, and that even though it was 1966, maybe women wouldn't be forever limited to support jobs.

TIP #1: DON'T SETTLE FOR LESS.

If things don't turn out as you planned, do what you need to do in the job you have, but keep your mind open for possibilities. Dwell on those as you continually seek the best job-fit for you.

Table of Contents

Chapter 1

Over the years I have seen the power of taking an unconditional relationship to life. I am surprised to have found a sort of willingness to show up for whatever life may offer and meet with it rather than wishing to edit and change the inevitable.

— Rachel Naomi Remen, M.D.
Kitchen Table Wisdom

STARTING BUSY

Whop! Things hit me when I least expect it. I was 34 and had settled in to life as a high school math teacher with three small children and a husband whose job in the Air Force kept us moving from state to state—a situation that had prevented me from doing graduate work anywhere, although I'd tried. Then in 1976 Ken was assigned to Dayton, Ohio, and it looked like we'd be there for a good two years. Along with the move and getting the kids established, I thought maybe this time I could go back to school. Maybe I could balance it all.

Computer science was hot then, a new field with lots of options; I'd chosen it over math for my graduate work just for that reason. I figured, who would hire a math teacher with no tenure who kept moving around? Ken worked with computers and saw a need for people with computer science background; he thought that field would give me more choices. I loved having options and knew we'd be moving again. Graduate work in computer science seemed a good fit.

Our budget was tight. While I did graduate work at the University of Dayton I decided to substitute teach math at the local high school, the one where I'd been nominated "Outstanding Teacher of the Year" back in 1966. Some of the teachers I'd worked with then were still there, so I knew I could teach the interesting classes. I found that I was the only accredited math instructor out of their 300 substitute teachers; when teachers knew they had to be out, they would ask for me. I got called often.

Life was full—I was a mother of three, a full-time grad student and a high school substitute at least two days a week. It required a lot of juggling, but I managed to get everything done!

I often went to the university as soon as my kids were in school—which was well before most college students were even up. That way I could go to the lab, start up the machines, put the tapes in and get the system up and running so I could use the machines without much competition. As I made friends in the lab, I learned that the University of Dayton claimed the honor of creating one of the first ten university computer science departments in the nation. The professor who had established the department in 1961 was still there and had arranged a deal for the school to purchase an IBM 360 in 1975, paying for it in 10 equal payments for the following 10 years.

Although the school had nine years to go on this agreement, one young professor had just purchased an entire room full of small Digital Equipment Corporation (DEC) PDP-11 computers for *less than the price of a one year payment on the big machine.* Technology was exploding so quickly that someone so innovative that he could found a computer science program just didn't recognize within only a few years how fast technology would become obsolete! This was the world I was stepping into.

While substituting and regularly teaching all the senior math classes for two years, I had earned the teachers' respect and felt like a valued member of the teaching staff. I was asked to teach full time while my friend and co-worker, Mr. Jarvis, had major surgery. Mr. Jarvis was so conscientious that he'd postponed his operation, worried about his high school seniors. But he was excited when he discovered that I could take his 5th-year math classes for five weeks while he recuperated. I knew the teachers well; many of them had been colleagues back in 1966 and I knew what was going on with them. So it was no surprise that I got wind of what sounded like a promising opportunity, and it came at a critical time for me.

The month I finished my MS in Computer Science--May of 1978—my four-year-old son broke his leg in a freak accident; job-hunting was going to have to wait until fall, even though I was anxious to put my new degree to good use. As high school wound down for the year, I learned from my colleagues that the computer science teacher had gone to California for the summer, hoping to find a job. If he accepted a job there, I assumed I could apply for his open position in Dayton.

Teaching would be easy with my children being in the school system, and I'd get summers off, so going back to teaching full-time seemed the right thing to do. I knew that in California a computer science teacher would get scooped up right away. I had high hopes for the fall.

In August, I got my call from the Assistant Superintendent. His computer science teacher had just quit. He didn't offer me much in salary, and that was no surprise. But he offered even less in incentive: he'd given away all the computer science classes to other math teachers—I'd get all general math and beginning classes. Nothing in my new field. *No* computer science.

Although this didn't sound very great, I knew the administrator couldn't find a replacement so late in the year, so I thought I'd help him out—with a bargain. I knew their system: with seven periods a day, teachers taught five classes, usually had one preparation period and monitored one study hall. I replied with my terms. I offered to teach four straight classes each day for half the salary and no benefits, so that I could be home with my kindergartener for one more year as he recovered from his broken leg, then the next year I'd work full time. I even was agreeing to teach the Assistant Superintendent's mindless general math classes, knowing he'd already given the computer science classes that I craved to people with seniority.

But Mr. Assistant Superintendent *refused* my offer, and then he dumped this lecture on me, telling me how teaching was a vocation and not an avocation, and he couldn't be pandering to different schedules for women who didn't really want to work.

As a wife and a mother of three, I'd balanced getting an MS in computer science with working as a substitute teacher almost every day while keeping all of my kids fed and in sports and school activities. The high school I just gave my soul to for two years had given me the shaft! A job was open in my field, I'd been their primary substitute in that field for two years, and I'd worked full time for them before that. What I should have asked myself was why I was willing to do so much for them for so little, but I didn't even think of that. Their computer-teaching position wasn't going to be mine, so what use was my Masters in Computer Science to me?

FROM TEACHER TO SOFTWARE ENGINEER

I panicked, as I wanted some experience in my new field and I hadn't been working all summer. I needn't have worried. I put my resume out to industry and had a computer science job in less than a week. On behalf of my son's need for healing time, I parried again when offered full-time work. This great little company, Systems Research Laboratories (SRL), then created a one-year job for me in which I'd work half-days and go back evenings twice a week, when my husband could watch our kids. At night I would teach a microprocessor course to my peers; after a year, the job would grow to full-time. I jumped at the offer.

Why were they willing to do anything to hire me? From a big-picture perspective, something broader was working in my favor.

Companies that had enough money were using huge computers. And had been for a while. Back when I was an undergraduate at Valparaiso University in 1965, the school offered a single course in computer technology and students wrote programs for a mainframe computer the size of a small bedroom. The mainframe only processed punched cards—and we weren't allowed to touch the computer at all.

I was familiar with the process. My Aunt Helen had made a living as a keypunch operator for Whirlpool Corporation, punching tan-colored Hollerith cards for software engineers, who wrote out their programs on coding sheets. Her job was to read the coding sheets and use her keypunch machine to type the program codes, putting corresponding holes into blank cards. Each 80-column card represented only one line of a program. Each column on the card ended up with some rectangular holes punched out—a "punch" represented a "1" and "no punch" represented a "0"—and the computer's card reader understood this language of 0s and 1s, from which it took instructions. A "job" consisted of a stack of maybe 200 cards, wrapped in a rubber band.

As undergraduates, we punched our cards and then had to hand our card decks to an operator who ran them into the card reader and returned them to us along with a report on greenbar paper, usually indicating that the program had failed. This had to be repeated indefinitely until the program worked.

I was using some room-sized mainframes to do my graduate work and operating them was still a time-consuming process, still punching my own cards and sharing the machine time. But because the university had that room full of mini-computers, the PDP-11s, we preferred to use them—they were plentiful and powerful.

Just coming into use were what we called "controllers," smaller, cheaper machines that were purchased to do a single, customizable task. More companies could afford them, and their use was spreading like wildfire. During my graduate work, however, no classes in microprocessors were even offered. In addition, other companies, like Intel, no longer just built memory chips and other "simple" chips; they began to build microprocessors – computer processors on a microchip.

Who would write the software for all of these proliferating devices? Companies were hiring anyone who knew even a little about programming, trying to keep up with the demand.

But there was more to my getting the job at SRL. Coupled with the big-picture forces, something on a smaller, more personal level came into play.

The manager who wanted me at SRL had engineers who knew nothing about microprocessors, and he needed someone to teach them. This man, Carl, had been a senior back in 1966 in the school where I was teaching. As the Math Club advisor, I had directed Carl and his friends as they came up with their crazy math projects. I'd loved what we had done together! Evidently the kids did, too.

So Carl had been ecstatic when he saw my name cross his desk in a job application. He decided that I'd teach his engineers at night, and then go in and work with them in the morning. If I didn't teach well enough, we were all in trouble, as we were building real things for the real world that had to work. What good training for *me!*

At the end of that frantic August week, Mr. Assistant Superintendent called me back, trying to weasel out an acceptance of my generous offer of the week before. Too late. By then I was moving on, looking forward to working at Systems Research Laboratories with a higher salary and flexible hours. After years in the school system, I had bridged myself into industry. There was no going back.

Systems Research Laboratories made it easy for me to get started: my hours were always flexible. That was important for a mother of three.

The work was very hard—actually just very challenging—and I loved it. I had no idea what I was doing at first, but not very many people in 1978 knew much about microprocessors, so my lack of knowledge of programming them wasn't much different than the rest of the world. I knew I could learn fast. And I did. This didn't mean I wasn't *way* out of my comfort zone, but I knew how to work hard, so I just took on the problems.

My teaching background had made me valuable to this little company. Since SRL wasn't big enough yet to hire the hotshots who were graduating from college in the top five percent, they got second tier people, kids who didn't know how to program microprocessors. But the company needed that skill. So SRL would have to train the people they had, and I was their means. I studied, taught everyone at night, and in the daytime we all worked together to put my teaching to use as we built the software the company needed.

Once we'd started doing good work and enough people knew what they were doing, my youngest son had finished kindergarten and I could work full-time, as I would for the rest of my career. But I still set my terms for working, and they just accepted me—the shortage of computer science programmers worked to my advantage. I worked 40-hour weeks by putting in 12-hour days Monday and Friday (often more than 12 hours, because I knew a babysitter was with the kids until Ken got home), but Tuesday, Wednesday and Thursday I worked from 7-2:30 p.m. I was home when the kids arrived on the bus, ran my carpools and took care of my kids. Flexible hours made it all work. Great life.

SRL GROWS A SOFTWARE ENGINEER

At Systems Research Labs I was older than most programmers—maybe that meant I seemed reliable and wise. So after successfully working with and teaching microprocessors, they gave me a project in trouble. I was to automate a manual system for General Electric (GE) in

Cincinnati that would drill holes in airplane engines. We had purchased Gandalf systems, and their modems managed the drilling of the holes, but we had to write the software to make the systems work. Up until then, technicians would "eyeball" where the hole should go and drill it, but computers would be more exact. We had to have user interface software on the computer for the operator to direct the work, and we had to create the software that would move the Gandalf machines that would drill the holes.

When I joined the project, the lone programmer had been writing code for six months, based on weekly visits during which GE technicians told the programmer exactly what a particular screen needed to look like so, they claimed, they could do their job. This request was totally irrelevant to what was now needed, but the technicians wanted to see a process that looked exactly like the job they currently were doing. For example, they wanted a lever on the top left of the screen, because they'd always had a lever to pull. When drilling the hole, they wanted the lever to come down, so they could "eyeball" how it was lining up— because they'd always done that.

Well, with less than a week to go on a six-month project, the programmer had produced an elaborate user interface so that the screen showed the right levers and buttons and things, but he had written no code to make the Gandalf machines actually *do* anything! It only *looked* impressive at that point. On top of that, the system was being built using little mini-computers, LSI11-03's, which had limited disc space and limited memory. All that disc space and memory had been filled with the user interface code, and there was *no room* to write any more code to make the system actually drill any holes. So I had to evaluate all of this with the other programmer, figure out what still needed to be done, and do it. But first all the FORTRAN code that had been written in those six months had to be re-written in assembly language in order to save enough space for the real work. Huge job, and a lot to test. Today we'd never have to do that. With current systems, memory and disc space are so cheap.

When I finished this project and it worked successfully, I was moved to another "project in trouble."

HARDWARE/SOFTWARE

I was assigned to another aviation project, this time to create a machine that could detect cracks in airplane wings around the bolts. Having a device like this would allow airline mechanics to *not have to remove the bolts* to check for cracks—the procedure that was used at that time, which, in itself, was damaging the wings! There had been a number of plane crashes around then due to cracks in the wings, so the government and the airlines were jointly focused on fixing the problem.

For this project I was assigned to Ed, a manager who had 11 children and a "wife at home where she belongs." Ed badly needed a good computer programmer. I fit the bill. But when I was assigned to him, Ed refused to accept me. He didn't want another woman. A previous programmer had gotten pregnant and left, and then he'd had to take a second female for this large project. Enough! He'd be darned if he was going to accept *another* one!

The powers that be wouldn't give him anyone else but me, however, so after a lot of haranguing, and the "promise" from my management that I'd stay on his project for a minimum of two years, Ed finally allowed me to work for him.

When I arrived, the team included a group of about eight men who comprised the hardware group, and one software engineer—Brenda—a young woman without a degree, who had first learned to write code from her uncle when she was in high school, and who had continued to try to teach herself. She was a cheap hire with a lot of limitations. Brenda was good, but she was timid—and mostly she was intimidated by Ed. She certainly never responded well to him!

When SRL's head of engineering added me to the project, making Ed's software team two women, Ed was grumpy. He took me, but he didn't have to like it. As far as Ed was concerned, Brenda and I had two strikes against us: we were female, and we were software engineers. Those facts alone already made us second-class citizens on his hardware-centric team, so we just didn't rate much of his time or attention.

What mattered more to us was to work with our hardware colleagues to solve the problem our team had been given. Together the team decided to use a laser. Sitting our little machine on the wing over the bolt, we could find the cracks around the bolt with the laser, send the results to a computer, and build reports of exactly what each wing's bolts looked like! Quite a breakthrough!

Writing the software to make the hardware work, Brenda and I had to work closely with the men on the hardware team. But the software wasn't running at all when I joined the group, and needless to say, Ed never cut me any slack. I was able to work well with Brenda, who couldn't see any light at the end of her tunnel, and who was glad to have someone help her out. The challenge was that any problems with the system were assumed to be software problems unless we could *prove* that the hardware was at fault. In the process or proving that, it wasn't very long before I became pretty good at understanding the hardware as well. This was a hint of things to come: for most of my career I would combine the processes of creating hardware and software programs that together made working systems.

A lot of software needed to be written for this project, and one lonely programmer hadn't been able to produce it fast enough. It was a miracle (or a combination of determination, smarts and dedication) that just the two of us rebuilt all the software, and got it working with the hardware as it should. But it was also the fact that Brenda and I never thought we *couldn't* do it. We just had to figure out how.

This was challenging work, and we did it with no tools. The debugger that we could have used had taken too much room in the memory, so Brenda had gotten rid of it. If software didn't make the hardware work correctly, we'd know it because we could hear motors grinding! Then we'd have to go *read our code* and figure out what we'd done wrong!

I didn't know it then, but these were the days I was developing my personal work practices. I began career-long habits of finding tools to make the programmers' job easier, and always focusing on working "smarter." This time with Brenda showed me that brute force wasn't always the best way to make improvements, even though we made software that worked.

In the meantime, and because this project was for the government, SRL had already bid and been awarded the contract for a more advanced Phase II version of what we were building in Phase I—in fact, the company was already bidding for Phase III work. Brenda and I built elegant software that made the Phase I system work flawlessly and—who would guess?—we built a fantastic first release. The product worked just as predicted and did the job that needed to be done. Phase I was everything the airlines and the government needed!

But had we built the first one too well…?

In my mind, as a taxpayer, we should have quit putting money into this project, and just let people use what was so needed by the airlines. It was an unqualified success; we could have called it "done." But since SRL had already been awarded the contract for Phase II, we then had to do it in color, and do it with seven angles of rotation. Now we couldn't just position the machine over the bolt—we had to be able to put the machine anywhere *near* the bolt, and make the machine move itself accurately into place! We added so many bells and whistles that the Phase II product became an elegant but cumbersome system to get up onto an airplane wing, with more chances for failure. Very whiz-bang, but I couldn't feel good about doing this forever. At the end of Phase II, when that product also worked well, I was not enamored of doing yet *another*, more souped-up version. But the government had been pleased with our efforts, and we'd already received dollars for Phase III.

So I was more than interested when the opportunity to manage another significant project opened up within the company—and they came looking for me.

SRL management had to get Ed to consider giving me up, a bit shy of my requisite two-year term that had been negotiated with him when I arrived. How did that go? They talked long and hard with him, but Ed continued being obstinate. So I went to Ed separately from management and said I was willing to stay with his project on one condition: I wanted to be project leader of the whole next project. Hardware and software both. All eight hardware engineers and two software engineers. I would take it over and be the project leader for Phase III.

Ed turned purple. A programmer managing the hardware people? Appalling! *Never!* He didn't believe a woman could do it. And he sure didn't believe a software engineer should be in charge; it was a hardware project! Software programmers did not manage projects. Certainly *women* didn't! Besides, what did I know about hardware?

Well, I replied, what did the others know about software? And I'd learned a lot about the hardware. In fact, while I was learning about the hardware, no hardware person was bothering to learn about the software!

By the end of our conversation Ed, who had *no* intention of letting me lead *any* project under him, let me leave to take on the project outside of his group.

I have to say in hindsight that it must have seemed easy to assign me, a woman, to a project in trouble. I didn't seem to care what I was getting into. I felt like we could develop software to do anything, and I hadn't experienced failure. I was expected to fix everything and make it work.

I was always intrigued by the challenges, and didn't realize that some jobs were so hard you shouldn't take them on. Even though I wasn't an experienced programmer, I knew how to get things done, so my first projects worked out. I was naïve and sometimes clueless in what I'd agree to take on, but from the outside it looked like I was fearless.

TIP #2: KEEP IT STEADY.

Take one task at a time and do it well. Or break the task into pieces until you can make it work. As you solve the problem before you, the next hurdle becomes easier. You're never in "over your head." You may just need to learn to hold your breath until you can surface.

Chapter 2

America is the greatest engine of innovation that has ever existed...a risk-taking culture with no stigma attached to trying and failing....and financial markets and a venture capital system that are unrivaled at taking new ideas and turning them into global products.

—Thomas L. Friedman, Pulitzer Prize winning columnist of *The New York Times*

FROM GOVERNMENT FUNDING TO CAPITALISM

Systems Research Labs had been building excellent plant management systems for years. Each system was handcrafted, then installed and painstakingly customized by a team of engineers sent onsite for that purpose. But following that model, by the time the team got the systems working exactly as the clients wanted them, SRL had lost its shirt on the projects. It was 1979 and the process was being repeated: another complete system had just been installed in a big company in Massachusetts. Customers loved our plant management systems, but not one had made SRL any money—much less broken even.

Enter Isaac, a software hot shot who knew the latest technology of the day: SADT. "Structured Analysis and Design Technique" described and clarified functions and relationships within a system. Applying SADT to a complex system yielded a diagram of small boxes, each with input lines coming in, output lines going out, and constraint notes on the top to specify how the boxes could be used. At each level, a box could be made up of more little boxes that further defined each task. Together, the full collection of boxes and lines described the original complex system in simpler terms of individual components and the flow of data between them.

Isaac was hired to create a *new* plant management system, using SADT charts and what was known from SRL's old systems. Each part of the new system would require a separate module, so all the hardware and software could be designed once, in pieces that worked together. Then, when a client called, SRL would go to the company's site and drop in the desired software modules, complete with our standard hardware and software. It would work beautifully and Systems Research Labs would get a check. Anything more the customer wanted would be customizing our system—and we could charge a lot extra for that. *Voilà,* a profitable business!

Modules were being made for all aspects of plant management: security, configuration management, maintenance dispatch, energy management, monitoring the opening and closing of doors (facility management)…the list went on. SRL would be able to meet the needs of customers—whatever they wanted—using the same hardware and clever software modules.

That was the plan.

With a few others, Isaac (a genius!) designed the complete software system with his SADT charts. Engineers could see that the output of one box was the input to another, and could note the constraints we had to design into each "box." They pored over pages and pages of the design, and it was beautiful. No, it was elegant.

When the design was done, Isaac was happy with it and assigned the boxes to individual engineers. If they all completed their tasks right and wrote working software code according to the described design, then

Isaac only had to put all of the boxes (i.e., all of the code) together, and everything would work.

Only it didn't!

Isaac had been unbelievably clever in his design, but he failed to confirm that the engineers who were implementing all of the boxes understood their assigned tasks, or the "inputs" they were getting from another box. Isaac trusted that everyone would do the right thing, and all the software engineers *thought* they were doing the right thing because the design was so thorough. They also thought they could direct the hardware because they knew clearly what the hardware engineers were building. But the whole didn't work, even though most of the individual boxes did. A case of the whole being less than the sum of its parts.

Irate management fired Isaac. But the problems ran deeper.

The project wasn't a government "cost plus" project, in which the government would pick up all costs of a project plus the computers and equipment needed. This was a project for profit—and SRL needed the money.

Could SRL cash in on the considerable man-hours and money invested by just putting the already-working modules up for sale?

Not an option. Unless they worked as a system, the modules couldn't be sold. And Systems Research Labs *had already sold this project to be delivered in months!* SRL had received money up front from the General Motors Oldsmobile Division, and SRL used the initial dollars to purchase the equipment and pay engineers' salaries. Company investment had been high: more than 20 software engineers had created some kind of code; the hardware engineers were about done with their hardware.

SRL needed to deliver the complete working product within nine months or we were out of money. Finished. Delivery deadline was approaching.

And nothing worked.

TAKING OVER A FAILED PROJECT

I had gained a solid reputation on my last couple of projects for Systems Research Labs, so when management started looking for a savior for this plant management system project, people in the company suggested me. I didn't think about why.

But I did think about not having to work with Ed on his now third airplane wing-crack detector.

In addition, the project looked interesting, and I was encouraged by management's faith in me. I agreed to take the job.

So with no salary increase and no changes in hours or benefits, new expectations of me were set: I was assigned a much harder task, required to manage 20 people instead of one, and had to deal with our customer breathing down the company's neck. In short, I was to fix everything.

What was I doing taking over this mess? I knew nothing about plant management systems, or about much of anything I was supposed to be doing—but I did know how to work with people.

LEARNING THE SPECIFICS

The project was laid out like this. I headed a team of 22 engineers. We reported to Sully, the head of software for the company, who assigned us to complete the software for this Plant Management System. Because SRL was matrix-managed, for the duration of this project I'd have two managers. Ron, a hardware engineer, headed not only the six-member hardware team, but also was the general manager for the whole project.

Our plant management system had three components: monitoring, analyzing and reporting. We needed a different set of computers—and separate, specific software—for each. We needed to develop and test so the three different components all worked together seamlessly.

For monitoring, our hardware engineers built some little custom computers to gather data out in a plant. The software for these computers—written in a strange but clever language called FORTH—could monitor situations, and tell if doors were opening, for example, or keep track of the energy that was being used. The FORTH software had been defined in SADT charts, but when I took over, the implementation wasn't making the hardware do what we wanted. If we didn't collect the right data, we wouldn't know how to manage this plant. (The problem could have been a hardware issue, but when I started we didn't know enough to be able to make such a claim.)

The second set of computers—Intel 8080s that we bought, not built—analyzed data they collected from the little computers. If the analysis indicated an action should occur, the 8080 had to determine the process for the action and make that action happen. For example, if data indicated the energy consumption in a part of the plant was too high, the action needed was to turn something off. The analysis would be: "Turn off the air conditioning in the manager's office, but only for two minutes, not three, or he will notice. If that doesn't adequately reduce energy consumption, then turn something else off, or many 'somethings' if need be. And then get them on again by the time they're critical."

We wrote assembly language software for the 8080 machines and though everything that needed to be done was in Isaac's SADT charts, these were hard, complicated tasks that all had to be made to work.

Each Intel 8080 computer gathered the data from the small machines, analyzed it and sent out task instructions to the small computers. Then the 8080 sent information on what was done to our third set of processors—a "big" DEC PDP-11 computer, which we also bought rather than built. Software written in FORTRAN for this PDP-11 machine generated reports for the company.

For security modules, we tracked every person who went through every door; for facility management we tracked doors opening and closing; for energy management we tracked where energy was being used.

Each morning after he'd gotten his coffee, Ron came into my office wanting a report about how the software was doing. Then he went out to work with his engineers to get the hardware done while I struggled to get the software done. He didn't know a lot about software, but he was smart. If I sounded like I knew what I was doing, and could explain to him just where we were, he just trusted that I would do this and get it all right.

I studied the designs until my head was swimming, poring over SADT charts day and night, when I wasn't carpooling my children and helping them with homework and making dinner or doing laundry for my family. But I was ready to manage the work.

I surveyed the engineers to determine what we had and where we were. I found that Ben and Sam, two senior designers who had worked with Isaac, knew at least as much about the system as Isaac had, and had installed a number of the "crafted" systems that Systems Research Labs had done in the past. They knew plant management; I knew they were our main salvation.

I put Ben in charge of all of the Intel 8080 software. This was the hardest part, as the 8080 read in all of the data and then acted on it. Ben and his 10 engineers had to write in assembly language, a low-level code that is harder to debug, and tricky. A lot of their work was writing drivers. Drivers are programs that operate, control, or regulate another device, and are usually written in a low-level language that lets them communicate directly with that device. I put Sam in charge of all of the report generation software on the PDP-11s. Sam had about 10 engineers writing software modules in FORTRAN.

I made sure that Sam and Ben didn't have to manage the people they worked with, but just managed the end product that we were producing. They were good at that. I directly managed the engineers who were writing the drivers that got the information from the little boxes running FORTH up to the Intel machines, but Ben stayed responsible for the goodness of the technology, and he made that software fly.

LEARNING THE POWER OF MY PEOPLE

I remember one of the really smart engineers who were working on drivers. I sat with him one afternoon as he explained what he had done. David was Chinese, and we went round and round discussing his code, because it wasn't making sense to me; I thought we had a language barrier, because sometimes his speech was hard for me to understand. After digging deeper, I realized that David had found errors in the hardware that he was working with; he knew enough to know why his software wasn't working. But David, because of his culture, would *not* point out to another engineer that engineer's mistakes! David just kept being his quiet, diplomatic self while we were failing for several days, hoping the hardware engineers would find their problems. David had known exactly why the system was failing, but could he point out another's mistake? Never. Fortunately he had me, and I had no problem pointing out the hardware mistakes to the hardware engineers after I studied them enough with David to know what I was really talking about. Hardware problems were fixed, the software worked well, and I took home the lesson that you don't just have to be able to communicate with people from other cultures, you have to *understand* their culture to optimize the results you wanted from them.

One young dedicated but uncommunicative engineer was writing all the FORTH code—the only engineer who knew how to do that. So I managed all the FORTH code with him directly. He really appreciated some other soul who took interest in his code, who would review it with him and who could help him perfect it.

The team focused not on completing *all* the modules that Isaac had initially designed, but only those which GM needed for the completed system on our due date. So a couple of the modules were just left unassigned. We did reviews, we talked to each other, and we argued with hardware about what worked and what didn't when we got bad data.

And every couple of weeks the three men who had procured this system for GM and who would be responsible for implementing it were sent down from Michigan to Dayton to see how we were doing. Since I was the head of software, it was my job to "explain it all to them" and show them parts of what was finished. I kept a blazer behind my door, and pulled it out when I heard the receptionist announce, "Bev, your customers are here." The trio made quite a delegation. One man, an Indian, was huge—over 6'5"; one was *very* short—no more than 5'2"; and the third insisted on wearing black shirts and white ties, so they looked like a mafia group dropping in. Fortunately, they were smart, and fun to work with, and so we worked well together.

If I took our Detroit guests to lunch, I paid from an expense account. But I was always the only woman, and in that day, a waitress just wouldn't give the check to the woman! So that always caused a lot of banter. If I drove them to the restaurant in my old Volvo, my boss felt it inappropriate: they should only see our customers' cars in our lot—and certainly shouldn't have to ride to lunch in a foreign car! When I told my boss I'd drive *his* car if it was the "appropriate" kind, or suggested that the company could buy me a new one, complaints finally slowed down. But I found my own managers would work the visits so we didn't have to go to lunch often.

On the other hand, I overheard the marketing guy telling his associates that they *should take me along* to describe our system for new business deals that were developing, because I was a woman. His reasoning? Women shouldn't know about the kind of stuff I knew about, so my doing the talking kept the interest of the customers; SRL was more likely to be remembered and to get the deal. So I was a freak and they wanted me along for that reason. After that, I was always bummed when they asked me to accompany them. First, helping marketing took me from my real work of developing software, but also then I always saw myself as the freak show that would keep the customers' interest— because who else had a woman describing plant management systems? Yech!

IS THERE A LIFE OUTSIDE OF WORK?

I was married and had children to raise, so I had to figure out how to do my job and still handle the rest of my life. My growing kids had soccer games, scouts and so many school activities—and I needed to be a part of all of their lives.

This proved to be workable in computer science, since my 22 engineers all had "flex time," defined as coming in to work as many hours as you can and always being there when you are most productive. No one worked 8 to 5. I arranged to work 7 a.m.-7 p.m. on Mondays and Fridays, having a sitter stay after school with my kids until my husband came home. Sometimes on those nights I went home, had dinner, and went back to work. The other three days I worked from 7 a.m. until 2:30 p.m. and then headed for home and car pools and all of the things that allowed me to be gone after school on Mondays and Fridays. It allowed me to be a mother. Our kids did okay unless I had to fly to Michigan for GM or one of our other customers.

I made some funny tradeoffs during this period. Having time with my kids was critical for me. My boss knew this, but wanted me to finish the project. So I showed him there were two ways to get to these monthly review meetings.

If my general manager and I flew to Detroit the night before, stayed in a hotel, and drove a rental car from Detroit to Ann Arbor, we could arrive by 8 the next morning, spend the day working, and return to Detroit for a late flight back. I wouldn't be home before midnight. But if we *rented a private plane* to take us up to Ann Arbor, we didn't leave until 6 a.m. on the day of the review. By 8 we were in the meeting, at 5 the pilot flew us back, and we were home by 6:30. I proved that the private plane with three of us aboard was cheaper than the commercial plane and rental car and hotels, so they let me do it. My personal reason was that I wasn't gone overnight for my kids.

In those days, just as today, women couldn't be superwomen. But a job that had my soul was my fate, and my kids and my husband took all of my other cycles. So I just let my children have all of the time that wasn't work time. Friends were friends I worked with. I knew neighbors, but I didn't have time to be much of a friend outside of work. Still, Ken and I lived richly. We were both doing things we loved.

LIFE ISN'T ALL FUN AND GAMES

Strange things happened as we worked through the project. Many of my engineers played computer games in 1979. When they couldn't think anymore, any simulation game on the computer let them relieve their tension. They would play their hearts out, then go back to work. I knew a favorite game was on all of our computers. The "fire" key for this game was the "3" key…. and that is the key that kept failing on all of our keyboards. When I'd send them in for repair, I'd tell my boss that these keyboards must be faulty for the "3" keys to be wearing out so much! But to me those frequent repairs were part of the cost of doing business. Managers had to keep their people productive, and this did it for mine.

Engineers who went out together for a drink after work would go to a place that had a new "machine" that allowed them to play Pac-Man for fun. This was in 1980, and video games were new. Some of the engineers got really good at Pac-Man! The few times I played the game with them, it was no contest; they were really great at manipulating the joysticks. I couldn't believe that engineers would focus such energy on another machine in their spare time! A young, single female engineer in my group got really good at Pac-Man and was by far the champion player. Eventually she got "tennis elbow" from playing the game so intensely and came to work with one arm bandaged, unable to type on her computer for a while.

The system we created had color touch screens for General Motors' maintenance dispatch systems. In 1979 and 1980, this was *cool* technology to be developing! Clients didn't have to be able to type to use our systems—they just touched the screens as indicated and what they wanted to happen, did. They could see in red where alarms were sounding, could locate all the maintenance men working on the shop floor and could touch on a map the place they needed a maintenance man to go. We used state of the art technology in what we did, and it paid off.

Eventually, we got the facility management, energy management and maintenance dispatch modules done and installed. We had to work out a few kinks. GM shop floor operators didn't accept computers in 1980, so we had to make the computers as invisible as possible. We needed to have things happen automatically for them without them touching the keyboard.

As an extra value, we had programmed the maintenance dispatch modules to work with color touch screens so none of our clients' customers had to type anything into a computer. For example, GM maintenance techs simply could pull an icon of a maintenance man over to an icon of a faulty assembly line area to assign a person to the problem. Boy, did my engineers like programming those systems! So we had some bennies for the engineers, too.

Years later I heard about Bohr bugs and Heisenberg bugs. Named for physicists Niels Bohr and Werner Heisenberg, these software bugs were classified for the kinds of failures they produced. A Bohr bug was defined and reproducible: you could study it, deduce the problem and fix it. Heisenberg bugs were uncertain, complicated and more frustrating: they disappeared or changed when you tried to study them. You could get a Heisenberg bug when you put two pieces of software together that each worked fine separately, but didn't work when you put them together, or when you put them with the hardware. And you had no idea where to begin to find out what was wrong. Well, the SRL Plant Management System software was a mass of Heisenberg bugs, and didn't we know it. But we broke everything up into manageable pieces and worked through it in painful detail until we made it all right.

Once this project was over, I never used SADT charts again in my career. I forgot about them. But whenever I tried to figure out my added value to a project, I always thought of it as my ability to get a team to focus on the important parts of what they were doing. In truth, my talent was to focus the *plan* that we were creating, ensuring that we understood what we were creating, and being able to pare a huge undertaking down to the essentials. Did those SADT boxes in my brain flow over to influence all I did in the future?

Looking back, I can't imagine that I agreed to make all of this work, but we did. It was fun and challenging to create a product that was good for my customers and made money for my company. It was an impressive system.

TIP #3: PLAY WELL WITH OTHERS.

As a manager, you cannot do everything yourself. Understand and utilize those around you; persuade them to take the actions you need them to take. Convince them that an idea is theirs!

Engineers are a quiet bunch and fellow managers are worlds unto themselves. Learn to design with the former and collaborate with the latter.

Chapter 3

A good manager is a very high-powered man, and we need a lot of him. But at bottom, he is a routineer; his aim is to make things go smoothly. No, for the wild places you need an innovator in charge, a man who likes to take risks, a heterodoxy if she is female – somebody that can meet wholly new problems in unholy new ways – you see?

—Poul Anderson
"Esau," *The Van Rijn Method*

SOLD ONCE, SOLD TWICE, SOLD AGAIN!

Systems Research Labs next sought to sell our plant management system into another division of GM, a process which was going well because of the success of that first delivery. But they were also touting it elsewhere. It was 1980.

Before our system was totally accepted at GM, SRL had agreed to put it into a pharmaceutical firm in Pennsylvania. Smith-Kline and French wanted the facility management module and others we'd completed. But they also wanted the security module—which we hadn't completed yet. In fact, we hadn't even started that work. The wonderful idea of dropping in the software and having it work perfectly was great, unless the "boxes" on the SADT charts back in our design book represented software that didn't exist! It was fortunate that we now just had one set of modules to create. But it was a big one.

The security challenge was in allowing only certain people access to certain doors, and not to others. All employees had access to the main entry door, but some had access to labs, others to office areas only, others to multiple places. In addition, the software had to be written to allow certain employees to change that access *daily* as required.

We needed cards that the employees could use to show for access. SRL chose to use a new system: we bought chips that went into the intended walls and also bought cards with chips in them. The chips to be embedded in the walls just needed installing in the appropriate places, and the cards arrived in a box, so to make them work together was thought to be "just a matter of programming."

When an employee of Smith-Kline held his or her issued card at the door, the card's chip sent a signal to hardware we'd put into the wall. If the card showed the right access code, the door opened and the employee would be let in—without ever having to insert the card into a slot. (This was why they wanted it in downtown Philadelphia! No slots meant no vandalism and an always-working system.)

Of course it was more complicated than that. Because some employees were allowed into some doors and not others, and Smith-Kline could change access for any person at any time, the software had to be made to accommodate changes in a very timely fashion. Even the big "chopping gates" in the lower level parking garage were run with cards that would raise or lower the gates. This software had to be written not only to make the access cards work correctly, but also to work seamlessly with the SRL system and its little data-gathering computers, the Intel 8080 systems that were issuing actions, and the PDP-11s that did reports for security.

I didn't have anyone on my team available to design all of this software, so I did it myself. Actually, it was a joy to read the manuals and figure out how to design the security system using these cards. I hadn't done any design since I'd started working on this plant management system, and I found I loved designing. I was terrified, as I was putting my neck on the chopping block for my software expertise—skills I'd only been using sporadically to review the work of others. But I had a junior programmer who could code a lot of it for me, so I did all the design and discovered that what I gave him was easily coded, so I was even a good designer! Managers don't always get that wonderful feedback, or get to do fun parts of projects. On later projects, when software had to be designed for work I was doing, I wasn't averse to grabbing a bit of the design work for myself. It became my reward for juggling all the balls that a manager juggles.

Early "failures" in beta testing did not come from our software designs, but rather were the results of things we didn't think of. For example, I watched a woman standing in front of an entrance wall searching and searching in her *big* purse for her card. Meanwhile through the side of her purse, the card had already sent seven—or even a dozen—requests for the door to open. And access had been granted *every one of those times* even though she never actually found the card!

In another unforeseen situation, access cards given to some executives in our initial tests turned out to have a signal identical to one that a local box store used for their merchandise security clips. We discovered this embarrassing little glitch when one senior executive was stopped and searched at the box store because their security thought he was filching merchandise, when he was just carrying our card in his wallet!

I continued to have to defy, dispute and defend often when there were hardware problems with the systems we were building; the prevailing attitude at the time was that it was always the software that needed correcting. In 1980 software people were "weenies" and hardware people ran the show. They felt their hardware was right, so you had to *prove* to them that their stuff wasn't working correctly—or just fix it in software, which they usually felt we should have done in the first place.

But the hardware-versus-software dynamic was beginning to shift. The industry was changing—for the benefit of small companies like SRL, and also for software engineers. In 1965, Gordon Moore, the co-founder of Intel Corporation, had made a statement that became Moore's Law. He predicted that computer processing performance would double every two years. The expanding complexity of the chips Intel was currently building attested to this. Customers delighted in the increased capabilities they were discovering, and used their processors for every purpose they could think up. But that meant customers depended more and more on software programmers to write code that made those processors do what was needed for business. And although software would not forever be the second cousin to hardware, it was my reality in 1980.

So I was in software and a female…why I was even left-handed! I had minority status up the wazoo. But I made myself fit. To do that, and so I could justify myself and prove my points, I always had to know exactly what I was doing. *That* made me a better engineer.

TIP #4: ENJOY THE GOOD STUFF.

Tell your body that this is the work you're built for, and revel in what you're doing. Help the body thrive, and absorb an attitude that says, *"This is good. I like what I'm doing."*

Chapter 4

I wanted a perfect ending. Now I've learned, the hard way, that some poems don't rhyme, and some stories don't have a clear beginning, middle, and end. Life is about not knowing, having to change, taking the moment and making the best of it, without knowing what's going to happen next. Delicious ambiguity.

—Gilda Radner, actress
and comedienne
It's Always Something

SAYONARA TO SRL

The software was working in the GM plant and at Smith-Kline, and we were testing modules that were going into a second division of GM. With several working versions of the Plant Management Systems that were functioning as planned, the company could now mix and match the modules and "drop them in place" any time they got a new order.

Then the Air Force gave Ken new orders: we were headed for a base outside of Boston, to arrive in July. Giving up my plant management software and my team would be difficult, but I was consoled by the confidence I had in my engineers--they were smart, they knew these systems, and they could work without me. So I organized a transition, said good-by, and got ready to move on. It was 1981.

Both Sully, my immediate boss, and Ron, my matrix manager, showed me that they appreciated my work. I had learned the most from Ron; he knew how to manage software and shared a lot of those secrets when we interacted. But for the rest of Systems Research Labs, I was just a departing employee, my work left behind.

In my exit interview, the VP of Human Resources said that she'd watched me and was crying when I left because I was the only woman at Systems Research Labs who had a chance at a VP position, and she'd been rooting for me. That surprised me! I didn't even know her. Did any other manager ever hint that what I did was worthy of a promotion? No, but I don't think I even thought of career opportunities at SRL. I just wanted them to give me experience and a chance to do exciting things. They did just that.

LIFE WHEN YOU NEED A NEW JOB

It hadn't been intimidating to get schoolteacher jobs. I had found new teaching positions in Ohio and Colorado, and part-time positions in Alabama and California; I knew how to do that. But now I had a different work history. I was going to have to interview now as a software engineer. That was brand new—and terrifying. Leaving a job I really liked didn't make it any easier. I wasn't really ready to do this job-hunting adventure, but I had to scout out the new territory.

Arranging to leave our kids with my parents, Ken and I scheduled a scouting trip in May, to last just five days. In that time we had to buy a house and I had to get a job.

While we were still in Dayton, our neighbor gave me the name of a contact working for a computer company in the Boston area. Right away I wrote this fellow, Adam, about my plan to interview in the area, and was pleasantly surprised when he called about ten o'clock one evening. First he grilled me for almost an hour, asking probing questions, then he asked when I would be in Boston. From the zeal he expressed on the phone I thought Adam would be fun to work with.

I started to consider the reality: few computer science majors existed; the job market was wide open. I began to feel a little more encouraged about the impending process.

Ken and I arrived in Massachusetts on a Sunday and spent the day looking around. Monday's first job interview was with a company that did what I'd been doing in Dayton, except they customized *all* their plant management systems. They loved me—I had a proven track record—and they made me an offer that morning before I left. I noticed no other women in the company on my visit.

That afternoon, for my second interview, I met with Adam and his group. Adam headed an engineering group of more than 500 people in Digital Equipment Corporation (DEC). He walked me in, introduced me to one of his engineering managers, and the interview was off and running. I got the sense that, based on Adam's phone conversation with me the previous week, he had pretty much told his engineering manager to find a place for me. I was handed a written offer before I left the building. Two job offers in one day! Oh, my!

Tuesday I had two more interviews, the afternoon one scheduled with another segment of DEC. By my fifth interview, on Wednesday morning, I'd glazed over, and decided it was my *last* one—after all, I had a house to buy, and two days to do it. With several job offers to consider, I knew I just needed to choose one and get on with life. I couldn't believe that in three days I had gone from being an anxious interviewee to being a confident soon-to-be new hire. Software engineers in 1981 could pick their positions, that was sure, and I found myself in the catbird seat. As a 37-year-old female with three years of solid software engineering experience, I was really being courted!

I did much soul-searching. Why I would build another plant management system when I'd already done that? Did I want "easy" or "challenging?" Should I build on my expertise, or start over with something new?

Finally I accepted Adam's job offer at Digital. I had used DEC's little LSI-11s and their PDP-11s in my last job, loved the company's current movement into smaller, cheaper computers and liked their software, too! A week after I accepted that job, I got an offer from the *other* Digital interview I'd had, offering a different amount of money. I found it surprising that a company as large as Digital Equipment Corporation had no common job screening, and that each part of the company could vie with each other for employees. I decided I was going to work for this company anyway.

I'd come a long way, baby, but felt like a neophyte again.

BEDFORD BOUND

With Dayton in the rearview mirror, we began our journey to our new home in July, arriving in Massachusetts with our little RV chock full of family belongings, three kids, two gerbils, a cat, a dog, and a car in tow behind us. Temporary living arrangements seemed straightforward. Staying at the campground on the Air Force Base in Bedford for the two weeks it would take to close on our house allowed Ken to drop off the kids (now 13, 11 and 8) at the lifeguarded pool while the two of us reported to work—he on site and I in New Hampshire. Digital had offered to pay for a month's stay in a hotel for us, but I couldn't figure out how to take advantage of this. Logistically it was a great plan—a beautiful hotel was right near my office building—but their pool had no lifeguards and I couldn't imagine leaving three kids alone all day for two weeks in a hotel room!

Days after we arrived, my husband was sent TDY—Air Force jargon for a temporary duty—meaning he had to go "somewhere else." I had to be up and out early each morning to get to work, about 20 miles from the base. With our plans trashed, voices piped up. From the kids:

"What about the pool?" And in my own head: "What about the supervision?"

Luckily, we were not the only family waiting for permanent housing. I found a wonderful woman who, while her house was being built, was staying with her two children in one of the Air Force mobile homes used to temporarily house officers at the edge of the campground. Ken and I quickly rented one of those units for our temporary quarters, parked our RV next to it, and arranged for this woman to watch our kids while I was gone at work. They had a TV; they had a huge campground in the woods on a military base; it was a pretty safe environment. Nonetheless, I'd race off to work and then race home each night, worried that they would get lost or hurt or whatever. Or that I would get lost *myself,* while coming back home along such unfamiliar roads.

After two hectic weeks of campground living, we closed on the house. I took a single day off work when the movers arrived; we finally got our family into our new home.

And that was the word of the day for us all—*new.*

The children seemed to do fine with their surroundings, suddenly having new house, neighborhood and schools to explore. So much was interesting, but also a bit scary for them. Having left all their friends, they needed more attention as they struggled to be accepted and learn their way around. I, too, was trying to adjust to the "new," whether it was the job and work environment, or setting up the home, or just navigating daily life in this different location. And since Ken seemed to spend a huge amount of time traveling for his new job, most of the challenges were mine to address—finding a new garage man, dentist, doctors, hairdresser, plumber. Things that are easy when you are settled in an area become huge when you have no support system. But we all managed to make it through each day.

My biggest faux pas the first week was during a search for a dentist, due to an emergency. One neighbor had left us cookies with a note to call if we needed anything, so I did. I introduced myself, thanked her for the cookies and said, "Do you know a good dentist in town?" I was a bit surprised at the chill in her tone when she answered, "Well, Dr. Stone is good at dentistry. I go to Dr. Parks. You can take your choice." Weeks later, I learned that her children all had the surname "Stone." How was I to have known that this lady had recently divorced one of the only two dentists in town?

Another error we made could have been called a "faux papa." Back in Dayton, our son had a male gerbil and, for a good while, we'd kept another that had belonged to a neighbor, because that boy's Mom wouldn't let him have a gerbil in their house. When we moved, the two kids decided the gerbil buddies should stay together, so both little critters came with us. By the time we'd settled in Massachusetts we were the surprised hosts to more than just those two original gerbils. (At one point some months later, in many separate cages, we had 17. We couldn't separate them fast enough or give them away fast enough. They multiplied like microprocessors—but didn't sell as well!)

We'd been in our house about three weeks when, on Labor Day weekend, my daughter fell down the basement stairs, landed on the concrete, and laid there, unconscious. I realized with alarm that I didn't even know where a hospital was or how to get her help!

In what seemed like hours, but was only minutes, she regained consciousness, but couldn't remember where she was, nor that we had moved to this new place. Everything was strange to her. I was able to get her to the car with the other two kids in tow and, somehow dredging a vague memory out of the fear, found the nearby army post. I pulled up to the gate and asked for hospital directions. On later reflection I realized that Ken and I had driven past the post when we were looking at houses back in May. I don't know how I remembered its actual location. I checked my daughter into the army hospital, kept the boys with me and called my hubby, who was—miraculously—at his office that holiday weekend. Even though panicked and cold inside, I realized that wherever I looked I was surrounded by only helpful people, and they took such good care of my little girl!

Emergencies tend to put the rest of life into perspective. Recovering from our scare, other challenges fell into place. In the coming months, I'd find the energy to join a computer committee for the school system, sing in the church choir, and live at soccer games and kids' events when I wasn't at work. We all were finding ways to stretch, grow, and fit into our new community.

TIP #5: BALANCE LIFE.

You'll do your job better if you can enjoy other creative outlets and let your work simmer in the back of your mind, getting inspiration from other things you do.

Chapter 5

I have wondered if I am trying to force a life.
While the life I lead may not match the
picture in my head, perhaps the one offered
me is just as full of joy, its pigments just as
bright, just not what I expected.

—Richard Paul Evans
The Locket

FIRST DAYS AT DIGITAL: ONE STEP AT A TIME

With a quarter-century of computer innovations to its credit, Digital
Equipment Corporation in 1981 was a dynamic, high-energy place to
work, and I was eager to get started. The company immersed me with
personal contrasts: I was 37 and coming from a small company to join
more than 100,000 employees, whose average age was 29. But
professionally I was already familiar with Digital computers, having
used some at SRL, and loved how easy they were to program. Before
arriving, I'd done some homework to see how DEC had innovated and
evolved.

Three years after its founding in 1957, Digital Equipment Corporation
delivered to a single client—Bolt, Beranek and Newman (BBN), a
computer consulting firm in Cambridge—a revolutionary product: the
world's first small interactive computer, their PDP-1. Named for its
primary capability as a Programmable Data Processor, the PDP series
hosted a number of popular versions. When the PDP-8 arrived in 1965,
it was the world's first mass-produced minicomputer.

Ten years later, Digital was releasing innovative products on almost an annual basis. The company's explosive growth was due to their mass production of inexpensive computers that were easy to use. DEC's first 16-bit computer, called the LSI-11, was introduced in 1975 and contained the first single-chip Large-Scale Integration processor. I used the LSI-11 in the late '70s, and loved this little machine. In fact, I loved all the computers in the PDP series.

In 1976 the 36-bit DECSYSTEM-20 was welcomed as the lowest-priced, general-purpose time-sharing system then on the market. Systems Research Labs had been too small a company to utilize a general-purpose time-shared system; the one-job-in-one-job-out nature of batch processing was sufficient for their needs. But the transformational technology of time-sharing, which allowed many users to concurrently interact with a computer, had not been lost on Digital's customers—Fortune 500 businesses and universities that knew how to make the most of time-sharing systems.

Back in my SRL days, I had become familiar with Digital products as a customer, particularly appreciating the PDP-11 and LSI-11for their operating systems, which made them much easier to program. (An operating system is a collection of software that manages the resources of the computer hardware and provides common services for computer programs.) Even though the programming itself was easier with these machines, I had to use a *different* operating system for *each* of my three projects at SRL. That's a high learning curve for programmers.

But Digital was innovating to change all that; in 1977, the first member of the VAX family (for Virtual Address eXtension), was sold, the VAX-11/780. Developed with 32-bit architecture, up from the standard 16 bits, DEC's VAX series—a number of hardware systems—was supported in 1978 by a new operating system, VMS, which had been created to run on *all* the VAX systems. (With this arrangement, VAX products would sell in record numbers for years.) Better yet for customers, all the applications built on this operating system could be used and depended upon for their business applications.

I arrived at Digital in 1981 never having programmed a VAX computer, but was placed in a group that would be building software for these machines and others. The company had just moved all the software engineers who were tied to the VMS operating system and all its related products just over the Massachusetts border to a lovely college-like campus with a brand new building on sloping, wooded New Hampshire hills. A beautiful place to work, and especially convenient for us engineers to work together. Over time, Digital would add two more buildings to this campus, and I worked in all three buildings.

Adding to the external environment, the internal atmosphere was even better. It was fun working at Digital, exciting to spend each day there. I couldn't wait to get to the office, and I couldn't wait to go to lunch to hear about what others were doing. Everyone was working on impressive software: separate groups focused on an artificial intelligence project, on building a new editor, or on creating "bulletin boards." These were not products that Digital sold or intended to sell— in fact, *no one* sold them. These cool products weren't available in the marketplace, they were just available for us to use if they made us more productive. It was like being a kid in a candy store.

Each day I found other tools that I could use to come up to speed on my work. I used a shell over the VMS operating system that made it look like UNIX, just because I could. I had a little program that helped me understand where I spent my time each day. I used tools to write programs and to test programs. I could network with engineers worldwide at Digital because about four years earlier, DEC had become the first computer company to connect to ARPAnet, so we already had a robust networking system. I could check on a bulletin board tool and see where people were going to dinner that night, or what I should be feeding my cat. So much software was available that people had just "dreamed up"—all there for the taking!

I'd asked to be hired as an "individual contributor," and not a manager. This was a very technical company, and I wanted to be valued for my technical expertise. To do that I believed I had to program with the rest of the engineers, and that I could always be a manager. I immediately found that I loved the technology, and being paid to design and build software was innately satisfying. I was assigned as a programmer to a team of four engineers to develop a new tool that would be used by software engineers. Our charter: to take an internal prototype called STEP (System to Track the Evolution of Programs) and create a commercial product. STEP provided a way to store many versions of source code very efficiently in a small space, and kept track of all the changes from revision to revision. Like a "library" of code you could check out when you needed it, then check back in, thus always being able to work on exactly the right version of the code.

I wasn't a strong programmer, having done little coding up to this point, but I'd read a lot of code—a circumstance that became something of an advantage for the team. Fortunately, since the coding on this project was almost complete when I came aboard, I coded some, but mostly built tests. My success at this stemmed from the fact that I came to the project late enough that I could ask a lot of good questions, but I knew enough to create automated tests that would answer the questions I concocted.

I'd taken the job stipulating that I could only spend 40 hours a week at the workplace. My boss accepted that, but neither of us realized how crazy this was going to be. Each day I'd arrive at work at 7a.m. and leave exactly at 4p.m. If traffic was good, I could be home in 12 minutes! Ken would get the kids out the door and onto the school bus each morning, and they would stay with a neighbor for one hour before I got home. I couldn't stretch that and take advantage of the neighbor, as I hardly knew her and wanted to remain friends. And I strongly felt that my children needed my time. The great thing was that this was 1981, and my company was a computer company. I was given a "terminal" to keep at home and, using phone lines, had access to the mainframes back at work (like having "servers" or having my work in the "cloud" in today's world). And "access" meant any time day or night. So when the kids dropped into bed, I could try one more compilation of my code before turning in, hoping to get a clean copy to work on the next day. I could read my email from friends who worked late into the night. The company had my soul; I was connected as many hours as I could find. But only 40 hours were physically spent in the office.

STEP was quite popular. As we were developing it, more than 500 projects within Digital were already using the product, those project engineers trusting it completely to never lose, damage or corrupt the code being working on. All those fellow employees depended on STEP to never change code from what they had shipped or code-frozen. We felt the pressure; our code needed to be bug-free.

Our product was good, but we were having trouble getting it finished for shipping. Upper management called me in and asked if I would "co-project-lead." Their reasoning was that the project leader, a great coder, was not as polished in project leadership, and could use my help. I knew I could do it along with my job.

Fortunately, Camden was more than a great coder. He was a smart and rational thinker, was generous to this woman who had been imposed upon him by management, and he wanted the best for the company. He was a gem to work with! Camden was quiet and unobtrusive, and just expected everyone to work as he did once they had an assignment. So my part of the project- leader role became following up with the others, making sure they understood what they were doing and got help if needed. In the process we discovered that all four of us were left-handed, so we made a fun fuss over that. Over the years I found that a disproportionate number of software engineers are left-handed!

Co-project-leading a team of four was easy after keeping track of more than 20 engineers at SRL each day, but it cut into my time as an individual contributor. I'd asked to be an individual contributor, since I didn't know this big Digital Equipment Corporation and how it worked, and I wanted to be accepted by working from the ground up. I was frustrated that, after so short a time, I already had been tapped for work that was outside a technical focus. But because managing came easy to me and those skills were needed, I agreed to take on those new chores.

Shortly thereafter, I was tasked with another side job— this one more technical. Management felt that the more eyes that looked at the technology, the better it could be. They instigated a "buddy" system, assigning people to review projects other than their own. I became a "buddy" on the BLISS compiler project. A compiler is a program which takes "source code" and transforms it into a target language which we usually call "object code." The object code enables a user to run the program on the computer.

My job as a "buddy" was to observe and participate in all the reviews and design meetings with the BLISS team. Digital sold this compiler as a product, but—almost more importantly—it was also used for *all* the company's projects, used by thousands of DEC employees programming around the world. The BLISS compiler allowed programmers to do things at a pretty low level, the way an assembly language did. I learned it and loved what it could do. I was the only female tied to this project, and one of their only two "buddies."

Following an intense review a few months after joining the complier project, I was walking down the hall discussing these technical decisions with the other buddy, one of the smartest engineers I'd ever met. Suddenly he stopped, shook his head, and said, "I don't believe this!" When I pressed him as to what he meant, he replied that he never would have believed he'd be working with a *woman*, much less discussing something this technical with one! It seemed like an uncomfortable realization for him. So even when I would think I was accepted as a peer, collaborating with the team and contributing to good technical conversations, I was reminded that I was an anomaly. Being female made it harder for me to be accepted for a good while, even in this big company.

MONEY AND MONIKERS

Our STEP project had increased by one additional college hire, making a group of three men and two women. We worked as a tight team, and just never slowed down in delivering product. Our code was elegantly designed and conquered a hard problem. When it went to market we knew it was going to be a fantastic tool for all software engineers using Digital Equipment hardware.

The year was 1982, and a product similar to STEP existed in the "free software" that shipped with the simple UNIX operating system, but wasn't available elsewhere in the industry. (Digital would offer their own version of the UNIX operating system to run on VAXes, but not until 1984.) When we were almost ready to ship STEP, my boss's boss appeared at my door, asking what I thought of the $15,000 price he and marketing had decided on. He was getting cold feet, suddenly afraid we might not sell any products, and he wondered if we should cut it to $8,000 each.

The cost per use went like this. If a customer like DuPont bought STEP for $15,000, the software would be installed on the company mainframe and everyone who had terminals connected to that mainframe could use it. That could be thousands of employees all using STEP for a single price. Today, Microsoft charges per person for their software products, but back then Digital didn't have that concept.

STEP would be the *first* tool Digital was making available for sale, so I made a quick gut calculation and figured that companies would highly value keeping their source code "safe" for all applications. In addition, STEP could efficiently store up to 3,000 versions of a piece of software in the space formerly needed for one version. The value to customers was clear: $15,000 was a great deal. I talked my boss into sticking with the original price, figuring we could always lower it if the product didn't take off. We never had to. STEP sold like hotcakes, consistently ranking among Digital's top 10 software products for the 13 years I was with the company.

But while the price was right, the name was wrong. Before it shipped it, the little marketing group tied to us decided our "STEP" could be confused with an education product with the same name. They researched…and researched. Each week we were closer to shipping, and they still didn't have a name for us. Problem was, once they came up with a name, we'd have to change the name STEP that had been hard-coded into a lot of the code—and everything would have to be tested again. I was getting upset; I knew what a job it would be to make those changes. I made a mental pledge: for later products, I always required that a product name be a variable, so any name could be put in at the last minute and it automatically would be replaced everywhere. Meantime the clock ticked on.

Finally, marketing came back to us, deciding to call the product CMS, for Code Management System. I was *so* exasperated. After lots of dollars, lots of research, and lots of whatever else they did, marketing chose CMS, which was the name of an operating system *at IBM*—the number one computer company in the United States and our biggest competitor! I couldn't believe they didn't know this.

We gave up the good name STEP because it was an education product, and instead were going to name it the same as our competitor's operating system? Yech! But it was too late to do another search, so in their wisdom, our marketing group tweaked the name to DEC/CMS. Confusion heaped upon indignity! The slash—being a ubiquitous character in software code, most often used to separate file names—was the worst symbol to use in the name of a product! But it had been decided. So we would have that dumb slash to deal with for all the years ahead. And a pretty undecipherable set of acronyms. But at least we had a name. And we could ship the product.

PRESENTING THE PRODUCT; SAVING THE BACON

The power of Digital Equipment Corporation's products was that they were great computers, in variable sizes that met various market needs, for a good price. Then suddenly DEC made better computers: VAXes. And once the VAX/VMS operating system was built to work on *all* the computers Digital made, a business's ability to customize its computers expanded rapidly. Users could buy a machine, build business applications on it, and then if they needed another bigger or smaller, our software would continue to work on any new machine. VAXclusters arrived in 1983, allowing many VAXes to be grouped together and continue to "act as one machine." (This provided a brief hiccup for marketing: If you had a huge cluster of VAXes operating as a unit, and STEP was running on one of the systems, was STEP then available to all engineers programming on all those machines for just the one price?) VAXclusters were so successful because Digital had incredible software, operating systems and compilers. But our job was to create more tools to make it even easier to develop software on our machines, thus making more people build applications for our machines and more customers buy them.

Watching the technology improve, I remembered the 8-bit architecture of the Intel 8080 I had used only a few short years earlier at SRL. Integers, memory addresses and other data units were at most eight bits wide, and the Central Processing Unit (CPU) was based on registers, address buses or data buses of that size. Those components grew to a 16-bit width in computers like the PDP-11, giving so much more capacity to the computer. And now VAX CPUs had 32 bits to work with! Remember Moore's Law: doubling capacity every two years! I was seeing it happen.

I was also seeing something else happen. In the early 1980s, more women were coming on board at Digital. There was still a huge discrepancy compared to male employees, but there was a noticeable attempt to hire more women. There were more women available in the marketplace to hire, and there was always a shortage of good software engineers, and both factors helped women in this field. I observed that life was easier for me than for the young women who were now beginning to work with me. I had quit working at 25 when my first child was born and stayed home for a handful of years until my youngest was four. That hiatus didn't hurt my career a bit—when I went back to work I just "started over" and with my experience I moved up the ladder quickly. But young women in general had begun to choose delaying having children. Waiting until they were 30 or older to have a first child usually meant that these women already had a good position in computer science projects, and at that point it becomes very hard to quit for a couple of years! So following maternity leave, women would have to go right back to work to keep up to speed in the industry, having to maintain the difficult, delicate balance between the needs of small children and a job with any significant responsibility. As I saw it, it often was hard enough being a working mother, but women who came to motherhood later in life had increased challenges in juggling their work and their home life.

Once DEC/CMS was available to ship, Camden and I were asked to go to Digital's Users' Group (DECUS) meeting to announce our new product. We built slide presentations and created the exact data we wanted the customers to know, sure that if they knew what we knew about the power of DEC/CMS, they'd buy it. Armed with our slides, info packets and posters we trouped off to the conference just to let customers know that our product existed. We had no market ads, no real marketing strategy. We just planned to talk it up, offer it for sale, and depend on "word of mouth."

The 1982 DECUS conference hosted at least 10,000 customers. When we arrived, we discovered that presentations were assigned to special interest groups: "operating systems" and "compilers." There were groups for FORTRAN, COBOL, VMS and UNIX. But nowhere was there a Tools group. Our DEC/CMS product ran on only two operating systems, VAX/VMS and Tops-20, and ran with eight compilers—and it was a tool that every FORTRAN, COBOL, Ada, PL/I, C, APL, or RPG developer would be excited to use. But we couldn't announce it in *every* compiler or OS group meeting; the conference officials limited us to only one announcement. So, where to announce it?

Camden and I made our rounds. The VMS OS group only wanted to talk about their operating system in their lectures. The small Tops 20 OS group had no interest in us. We had eliminated each compiler group. Finally, we learned that conference officials had put us in the UNIX OS group, because UNIX packaged some tools in its operating system! The UNIX group was appalled at this arrangement, particularly when they discovered that Digital was going to *charge* for this product, as the UNIX guys were part of the "free software" contingent. And so we were misfits—with a lot of work to do.

I sidled up to the head of the VAX/VMS group, negotiating the best I could to get them to announce us before every presentation in their group, and point all their people to come to our "UNIX presentations." I bandied with the UNIX people who were just out to give us a hard time. I went to the compiler groups and made little announcements asking people to go "see something new that Digital had to offer," even though they would have to go to the UNIX track to find us. What a week of being a salesman instead of an engineer!

Because this was my inaugural visit to DECUS, by later in the week when I had a couple of hours to myself I decided to attend some classes. I found a great one on security being offered by a customer, not a company employee. The speaker was articulate and knowledgeable. Fascinated, I was taking copious notes when I realized he was showing his audience of about 300 how to break into any computer connected to a network, and get into the nodes on that network! OMG! My head started spinning.

I knew that downstairs was a huge room full of Digital computers available for customers to try products on, and every one of them was networked back to our computers in New England, which at that moment were being used not only to develop all the software of the company, but also to do all the work of the company! Where would these customers try their newly found knowledge? On those available computers! And in less than 30 minutes, with the data this customer was giving his audience, all 300 curious customers could go down and "try this hacking" on the big computers downstairs, possibly bringing down the company, and certainly causing havoc to every group represented at the conference! (Digital attendees all connected back to our home computers to download software, and check on things, etc.)

I *had* to keep taking notes! But I also had to tell someone what was happening. I stayed a little longer, just to get the gist of what we were being trained to do. Then I ran downstairs. I knew the name of *one* security engineer in the operating system group, and I raced about, asking for him. Here I was, new to the company, running around looking deranged, and hollering for Zeke. Oh, boy! To my great luck and good fortune, Zeke was down there working on something, and he gave me the time of day.

In between my ranting and referring to my notes and giving a pretty good description of how users were about to hack our resources, Zeke began to realize the gravity of the situation. "How much time do we have before that session is over?" Maybe 20 minutes at most, I thought, counting the time it would take for those engineers to come down the stairs and attack our computers.

Zeke started hollering for helpers. He'd call someone, order them to do something, then grab someone else and get them busy. I was captivated watching Zeke, a quiet, mild-mannered engineer, suddenly become a focused commander of forces doing something so stressful and very difficult. Within fifteen minutes all network connections in that room were disabled. Within about seventeen minutes our horde of hackers had descended upon us, finding machines and trying to see what they could do. Hundreds of attempts to access our computer networks were made in the next few hours. All failed. The company was safe. At the end of the day, I was happy to just get out of there! But I'd made a friend in Zeke, and in coming years I'd seek him out with security questions, and he would always talk to me. From that point on Digital did not connect machines on the conference display floor to the company's home computers—except at night when the machine room was closed to the public.

In spite of having no special interest group to leverage us, our new tool that we knew customers would love was well received. People clamored for the DEC/CMS; it sold well and kept selling. Eventually DECUS created a "TOOLS" group at the conference to showcase this and other products that my group produced along with our compilers. Within five years I was well known and comfortable at these events, presenting products users were ecstatic about—I even began to enjoy going to user group meetings!

ENHANCING THE PRODUCT; LEARNING SOME LESSONS

So now we had DEC/CMS Version 1 selling as a great product, but it was somewhat slow. We stored all the information we needed to do our work in a bunch of files, but each file had to be opened at least once as we performed our operations—not an efficient system. For Version 2 we planned to build a database for information storage instead of using files, and we added our fifth team member as we started working on that.

J.S. was a smart kid, right out of college, and he didn't know that some things were too hard to do. By this time I was the sole project leader and I assigned this new kid to build us a database. Yes, other companies had huge teams of database developers, but our project had one—J.S. was 22, skinny, young looking and a little green. The only way J.S. could keep it all straight was for me to let him stay at his Cambridge apartment on Fridays and lay out the whole database plan across his dining room floor to check for discrepancies. I was *so* glad he and his wife were still "too early in their married life" to buy any dining room furniture until after we finished Version 2 and had the database complete!

DEC/CMS Version 2 was quite versatile, and our developers expected it to be used the way it was intended, through the graphical user interface (GUI). But sometimes products can be more effective if one product is allowed to call another directly, and bypass the GUI. For that we created an application programmers interface (API), which allowed calling directly into the product. But when customers happily discovered they could make direct, low-level calls to our product from their own products, it meant we had to develop hundreds of customized ways for them to do that.

Complicating the task, our customers often didn't just perform the standard call-in. Not knowing how to use our API, they called DEC/CMS from inside their products in strange ways, usually wrong ways, and made mistakes, some so bad as to crash the whole thing, theirs and ours. We'd have to spend hours figuring out why it didn't work for them: was the customer wrong in how they used our product or did we really have a problem with our software code? We'd created a *very* flexible product at the cost of soaring increases in customer service requests. To offset the problem, we began to educate our users and eliminate useless errors. We provided guidelines for using the API, and installed run-time checks immediately inside it to screen out (stupidly) wrong user calls.

My biggest practical lesson from this experience was, for the rest of my days, to build into my product the *fewest* number of calls that I could— certainly never hundreds! Few things are more frustrating for software engineers than to have to prove their code doesn't need fixing— especially when that code had been so thoroughly tested to get it into a product in the first place.

I also learned a personal lesson: that if you truly want to accomplish something, and you figure out exactly what that is, and you have smart people to work with, you will get it done. I would come to understand this much better later in my career. An engineer isn't asked if something can be built; nor does an engineer ask if it can be done. An engineer hears from customers or from marketing or from his boss that a certain thing must be created. He is assigned to the team that is to do it. Then the questions begin: How can this be done? What is the optimal way to complete it? And if that just isn't feasible, what is an acceptable alternative? But there is no question that it will be accomplished. Once I developed this mentality, my strength was to focus the design in on what was doable. We'd ignore what might sidetrack others for days or years. We'd focus on the task at hand, and find the best ideas that we thought could be implemented. And then we'd start running to make it happen. We had five engineers on this project, most pretty junior except for Camden, and we did some phenomenal things.

MORE PRODUCTS, NEW DIRECTION

On a separate level, the DEC/CMS product already had been generating a life for itself beyond the company. Before I arrived at Digital, an engineer on the project had taken the DEC/CMS code with him to another company. This behavior seemed to be okay in those days—to just walk out with proprietary code and never be penalized for taking it. That company used it for a number of years. At some point the product was spun out and put on many platforms, and so whenever our current, official version of DEC/CMS ran up against the products of competitors in the field, or similar capabilities on competitors' products, we found ourselves usually encountering something based on our product or a variation of it. We could be proud that, under our own name or not, we had owned the industry in providing the best source code management and revision product to be had!

I was still feeling strange in this job because I didn't ever interact directly with any hardware. The entire building of engineers in New Hampshire wrote software, tested it and sold it. We trusted the operating system under us to remain stable, and that's all we needed.

The OS group talked to the hardware teams, and built drivers to optimize the hardware. We built to that one operating system. We sold software without ever talking to or even *seeing* the hardware groups that built the computers that we ran our products on. Occasionally I would go into the computer room and load tapes, etc., just to make machines *do* something! I was used to having time-consuming and difficult battles with my hardware groups, but with this standard OS, I never needed to do that. This was *very* cost-effective for Digital.

By the time we were thinking about Version 3 of DEC/CMS, I was asked to supervise the group, continuing to drive DEC/CMS, but also working on creating an "automated test manager" product, and getting that built and to market. An Automated Test Manager was another tool for software developers, allowing them to write tests, put them smartly into a collection, and then run collections of tests, individual tests, or mixtures at any time. Developers could also time-stamp tests, and have groups of tests that needed to be run daily, weekly, or just before freezing their code for shipping. This tool made handling all these options easy for the developer. We struggled hard building the base for this product, but the product was very easy to use.

When we were ready to ship, marketing looked at this, the second tool in a collection for software developers, and wanted to name it DTM, for DEC Test Manager. Then they saw our weird DEC/CMS name, and decided that the new tool would have to be DEC/DTM. But *that* was redundant; they said no one could understand these crazy acronyms and we should be using *words*. Having learned our marketing lesson long ago, this time I had ensured that our engineers had put in a generic code name variable for the name of each product. We just had to change one name at the beginning of the code and automatically that name would appear in all the right places in the code. Good thing, because again the name was argued about until we were ready to ship. Finally, when "DEC/Test Manager" was chosen we accepted it, coded it in and shipped the product.

And we kept on going. Our group's third tool, DEC/MMS, was used to do automatic builds of software, which could then be replicated. I got to announce this product at the next DECUS, and could see that developers were hungry for tools in addition to their compilers. They even tolerated our strange product names and remembered the names as they ordered our products. But when we announced the DEC/MMS, and had three tools selling, Digital Equipment Corporation had officially moved into the software tools business.

TIP #6: BE YOUR OWN CHEERLEADER; TOOT YOUR OWN HORN.

Sometimes fantastic work goes unnoticed. Push your products, ideas, thoughts and inspirations. Sometimes you can push quietly, through your code or designs, but other times you need to do whatever it takes to be noticed! If company "fires" get all the attention and you have a group that runs like a well-oiled machine, you have to publicize yourselves.

Chapter 6

The one thing all famous authors, world class athletes, business tycoons, singers, actors, and celebrated achievers in any field have in common is that they all began their journeys when they were none of these things.

—Mike Dooley, former international tax consultant turned entrepreneur; author of *Choose Them Wisely: Thoughts Become Things!*

RE-GROUPING

When we built the first version (V1) of DEC/CMS, I was part of a group called Methods and Tools, which included compilers as part of those tools. Because there were other compiler groups in the company, management decided to bring them all into our building and reorganize them as two compiler groups.

The DEC/CMS team became part of the incoming compiler group renamed Commercial Languages and Tools, because it included the COBOL compiler and other business compilers, like RPG. The other half of Methods and Tools was renamed Technical Languages and Environments and included the FORTRAN, C, PI/I and Ada compilers.

The first tool this second group wanted to build was a Language-Sensitive Editor. And they'd been hard at work on it.

Ambitions were high for this product: it was being designed to provide help with program composition and to present error messages detected by the compilers. The challenge for the group was that Digital by that time had eight major compilers—one for each of the most popular programming languages--and the editor had to work with all the languages and their compilers. (Had we just built the editor for a couple languages, say FORTRAN and Ada, it would have been much easier.) Since it was a challenge to get this multi-lingual Language-Sensitive Editor perfected, finished and shipped, and because there were ideas for several other such tools, the manager of this group wanted to hire a separate manager to get it all off the ground. And since the manager of this group was my former manager in Methods and Tools, I was hired into the Technical Languages and Environments group to supervise the building of the Language-Sensitive Editor. I refused to take the job unless I had some compilers to supervise, too. I knew that it was really a compiler group, and that if I didn't have a compiler to supervise, I'd never be seen as "technical" in this group. My old manager not only allowed me to do this, he "loaded me up." Suddenly, like the multi-tasking language editor we were creating, I found myself supervising a handful of separate groups.

I was put in charge of seven products: the C Compiler (at that time seen as a UNIX compiler that ran on VMS), the PL/I compiler (for "Programming Language 1," but written as Roman numeral "I"—a distinction that has its own long history), APL (an interpreter), the VAX Debugger, plus three beginning "tool" projects: the Language-Sensitive Editor, the Source Code Analyzer, and the Program Design Language. We had very smart people doing all of this creating, and I loved working on these projects.

TOOLS BY ANY OTHER NAME

My skills in focusing technology and producing good products coupled with the skills of an outstanding project leader for the Language-Sensitive Editor allowed the two of us to estimate how quickly the project could be finished. Management wanted it done yesterday and the project leader thought it would take at least a year. The group began with lots of work, added lots of technical sessions, lots of excitement plus a *lot* of testing (using the DEC/Test Manager, of course!)—and we had a product. Testing it with eight compilers almost did us in, but the DEC/Test Manager rose to the challenge. When this editor finally shipped, it immediately sold plenty of product. Then we knew we could go to the DECUS conference and announce it within the newly-formed "Tools" group, which had been created once DEC/CMS had become well known. All our 35,000 customers at the conference would hear about it and be ready to buy it! Furthermore, I was excited that marketing had seen the light—we'd be announcing a "Language-Sensitive Editor," not DEC/LSE, or DEC-slash anything!

I was proud of this product, since it made coding so much easier and made it so easy to write good code. Our Language-Sensitive Editor anticipated most of what a software engineer needed as he or she wrote code—like a template, it filled in the spaces or helped engineers keep from forgetting the necessary items in a line of code that would make the code correct.

At the end of the first year, thousands of customers had reported only four bugs in this product, a record for any product at Digital. For that achievement I credit the diligence of the project leader, who simply tended his product that carefully. The code written by engineers on his project had to pass his close inspection before any changes went into his product, and it was flawless! In my entire career I can remember about five engineers who had the talent to just deliver *correct* code, and they were worth their weight in gold. In future years I'd be smart enough to use such engineers wisely, but in those days I just appreciated that this one was at DEC and working with me!

I appreciated, too, being so fortunate to work in a company at the forefront of the computer revolution, and having networking capability so early on. From my first days at the company I enjoyed being able to put a specification on the company's internal "internet" just before I left at night and have someone in Australia give me feedback before morning! Four years later, working at Digital was still exciting. In 1985 the company registered the fifth Internet domain name, dec.com.

After the Language-Sensitive Editor was completed, it was followed by our Source Code Analyzer tool, which gave programmers a way to look at their source code as they were working on it and get analysis of what it was doing and how well it was doing it. In addition, our Program Design Language tool made life much easier for the coder. Traditionally, once you wrote code, you had to compile it and it had to be all "correct" for the compiler to be able to handle it. Our Program Design Language tool allowed coders to "free-flow" the writing of their code and leave unfinished areas—code could be "compiled" without having all of its guts—and coders could still get the gist of what they were doing and whether it would work. Outlining tasks for an overview became easy; "details" that finished the assignment could come later.

Customers, once these products shipped, called them "LSE" and "SCA" and "PDL" to the deep chagrin of marketing. But we engineers just built these products, knew they were good, and continued to "not care" what they were named.

I was able to do something about these weird names finally, because I was able to create a set of six tools—my old DEC/CMS product, DEC/MMS, DEC/Test Manager, the Language Sensitive Editor, the Source Code Analyzer, and the Program Design Language Tool. We put them together in one big collection called "VAXset." I thought that bundling these tools would have been an easy thing, but it did require a little work inside each tool. My colleague, who now had taken over

DEC/CMS and the DEC/Test Manager, really didn't want any more on his plate, so he initially refused to do the bundling. DEC/MMS was under a third manager, so I had to work with both managers to make the VAXset happen. I realized that some managers weren't interested in possibilities for the company (like the ease in which customers could buy VAXset), they just did what they were assigned to do and that was enough for them to try.

Finally, we got the VAXset package coded, packaged and shipped. Our customers were happy to buy the bigger package and we gave them a bit of a deal: six tools for the price of five separately. The VAXset was so successful that it was on the list of the top ten-selling products at Digital Equipment Corporation for all of the remaining years I was with the company.

CRAFTING COMPILERS

PL/I was also fun to supervise. Len and Lana, the two hotshot engineers assigned to the project, were right out of college, but very smart. I decided that I'd work the specifications and develop the plans for this compiler, and assign these two the work itself. They both did so well that we had quite a strong compiler. Lana would carefully design and code her assignment, test it carefully, and then go into the shipping product, retrieve the documentation, and change every bit of the documentation so that it matched the new functionality that had been implemented. The technical writers loved her, as their work was done!

Once when the three of us went out to lunch for a celebration, Len expressed concern that we were all going to the restaurant in the same car. "What would happen to PL/I," he worried, "If we were in an accident and something happened to all of us?" From that point on we considered driving in separate cars, so that our compiler would always be taken care of.

Because Len and Lana were junior, I took it upon myself to be the Digital Equipment Corporation representative to the PL/I Standards Committee, a national group that defined the PL/I language standards by which each company had to build its product. PL/I initially was a humongous language created by IBM. Later, when the standards committee was formed for PL/I, the group believed the original language was too big to implement, so they created a PL/I subset that then came to serve as the standard. Annually, the committee could choose to add to that subset-standard either new things or more of IBM's language.

The G-Subset, as it was called, was pretty good and over the years most companies used that standard and created their own PL/I compilers, satisfied that what they had crafted was working well enough for their companies' purposes. I had gone to several PL/I Standards Committee meetings before I realized that most companies did not *want* to enhance their PL/I compiler—Digital was no exception; we had a solid, competent PL/I G Subset offering—and implementing each change the committee made was expensive. So at the next annual national meeting I requested that we as the PL/I Standards Committee call our work "finished" and abolish ourselves. Heresy! Engineers love a compiler the more they work with it and on it, and these guys all loved the compiler (as I did), and we were on the *national* standards committee and it didn't seem *right* to abolish ourselves! But after a lot of discussions, we did just that. I was so proud of this group of senior engineers who had the guts to say "enough." We froze the standard as it was, and it has stood as the basis for PL/I for perpetuity.

I also supervised the APL (an interpreter, but similar to a compiler) team, consisting of one full-time senior engineer (Rick) and one or two good "part-time" interns. Rick was the project leader, the designer, the coder, the marketer, the one we sent to our User Society forum to represent us. He and I were the soul and heart of this product. IBM had created the APL interpreter years earlier and by this time supported it with a team of 136 people in engineering development, coding, testing, customer support, marketing and sales. We felt that our APL team of one full-timer, two part-timers and me supervising about one-seventh of my time were doing more than just okay for Digital in providing a great product with new versions, which continued to sell well.

C was the language of choice for UNIX systems, and so to round out our offerings on VAX/VMS we created a VAX/C compiler that met the American National Standards Institute (ANSI) standard. Project Leader Bret had led the team to build this very efficient C compiler for VAX/VMS into one of the best in the nation. Bret was Digital's representative to the C Standards Committee, and his intelligent input continually influenced that group to make very elegant changes, enhancing the power of the C compiler. Bret's excellent ideas and suggestions to the ANSI Standards meetings were then quickly and easily embedded into the VAX/C C compiler when he returned from a Standards Meeting. We were always the first with changes, and we always touted a world class C compiler. Again, I supervised a small team for this work. I didn't get much recognition among my peers for supervising the C compiler, because the general opinion of the day was that "the good compilers were FORTRAN and Ada" and COBOL was the business compiler, so who would use C? But I loved this compiler, too; we gave it the attention it deserved. Eventually the C compiler became the one that carried the industry forward.

We also wanted to build a C++ compiler. A fascinating Digital group in Seattle was building compilers to use with the NT operating system being built by Microsoft, which would then be tested and used on Digital machines. I spent a good deal of time working with the Seattle group, doing some wonderful "trading of technology" and sharing the workload. Even though our group wanted to build a C++ compiler, the fact that one had already been created in Seattle brought us to the decision to let that group own this compiler development. This was my time to learn the importance of getting the best results for the company, even if my group didn't get to do the work.

By now I was managing about 80 engineers building this mixture of products, and I had to worry about the strength of my teams to do the challenging work we were asking them to do. I was still hiring engineers, constantly keeping an eye out for strong, creative people who would fit into my group. When one particular engineer would ask at every DECUS meeting for many, many enhancements to the Pascal compiler, we suggested that he come to DEC and implement those enhancements himself. He did come and was hired. He spent the rest of his career building the Pascal compiler. When I visited the Wang Institute of Graduate Studies in nearby Tyngsboro, Massachusetts, I met the head of the department, who was teaching his students to build interactive compilers. I challenged this innovative professor to come into business and use his techniques to make our industry-leading compilers interactive. When he joined us, Digital won because he contributed so much to the strength of our compilers. I was always watching for stars that we could add into our "firmament."

CHALLENGES WITH THE DEBUGGER

I was responsible for the VAX Debugger, and although this tool shipped as part of the operating system, it was more aligned with language functions, as it directly impacted code-writing and corrections. All our programmers needed to "debug" their applications with this tool, and our group was charged with maintaining and enhancing it. Initially, the VAX Debugger utilized a command-line interface resulting in long, successive lines of text. Everyone today has seen something like http://www.google.com/. Well, a debugger had to say http://www.debugger.com/x/y/z or whatever else the programmer wanted to have the debugger do, and writing all the instructions out was tedious but expected by users. That interface allowed the programmer more concise control and power in the interface process. The key to accessing the wonders of the tool was in knowing what to fill in for the "x" variables in a command string that began: "debug/x/x/x/x/x." But mostly only advanced users held that key, or perhaps, were the only ones who had the patience to stick with the attempt.

Around this time, pull-down menus and graphical interfaces were beginning to be seen as preferable to just "giving written commands." Modifying all our compilers and tools to have a graphical user interface (GUI) made computers significantly easier to use. All applications running on the VAX/VMS were going to need to comply with this emerging shift in technology. First the DECwindows portion of the operating system would complete its core work, then each application's interface could be modified to take advantage of the new graphical interface capabilities.

Digital's VAX Debugger needed to be at the forefront of this move and one of my early tasks was to make that process go smoothly. Since we shipped the debugger with the operating system, it seemed important that the debugger have the "new look" as quickly as possible so it would ship with its new GUI when the operating system version made the capabilities available to all. This turned out to be great fun for me, as I had the chance to let my creative juices flow. Back in the early 1980s an engineer in our group had been single-handedly creating a word processor with a totally graphical interface. At the time management didn't buy it, but I'd loved the idea. I studied advanced graphics courses at night and became addicted to seeing things visually and making products easier to use. The field of graphics was compelling to me, and continued to be for a good portion of my career.

When we made the debugger work with DECwindows, customers could pull down a menu of choices to take advantage of, and the power of our debugger was even more obvious. However, the first version of the VAX Debugger that worked with DECwindows implemented only *half* its functionality—and users *still* thought it was fantastic! This fact led me to understand that unless a product is easy to use, customers will not utilize all of those neat features anyway. From that point on, I was careful to monitor how much we implemented for any tool.

The debugger gave me the greatest challenge in 1987/88 when renowned hacker Kevin M. gained access to the VMS operating system, and the system developers were able to trace the problem to the debugger. Normally while it was working, the debugger for a short time would go into "kernel" mode, a necessary condition (predicated on the assumption that safe, trusted software is being run) that allows the computer complete control to execute any instructions. During that short period of kernel-mode opportunity, this hacker was able to take over and do damage to our system—and worse yet, the systems of our customers.

One day a senior VMS developer came to me, ushered me into a room and closed the door. He revealed what his group had discovered and gave me a few hints why they thought this was happening. My job was to surreptitiously pull engineers from their directed assignments on the debugger to fix this "hole" and fix it quickly and quietly so that no one else in the universe knew it was a flaw in our system.

The biggest challenge was that I was sworn to secrecy, not permitted to tell even my supervisors that we were working on this, until the problem was fixed and shipped. For about eight days, in meetings when my boss thought my people were working on other projects, five of my group were *really* working on this fix, and not doing what I was reporting that they were doing. I met with my accomplices three or four times a day and we quickly fixed the problem. I took the fix to the VMS group, who sent out an update. But while and until customers had installed their system updates, our little band of co-conspirators just went back to work and struggled to catch up on more than a week's worth of lost time, telling no one what we had done. A second time in my career, years later, when I managed the Alpha VMS operating system group, Kevin M. was again able to break into our systems and make us even more vulnerable. But then I knew just how to attack the problem, putting to good use my experience with the VAX Debugger breach.

I could have stayed in this Compiler and Tool group for the rest of my life; I loved the work we were doing and the smart people I was working with. I'd already been at DEC nine years, moving from Individual Contributor to Supervisor to Engineering Manager. I had thrived with intelligent and cooperative peers, and smart supervisors and their managers, who generously encouraged and supported us in creating innovative products that improved the lives of our fellow engineers—and thus were good for our business. But at some point my boss's boss's boss (Adam, the man instrumental to my being hired in the first place) called me in and said Digital was in dire need of a strong manager in the "Office Systems" area. They'd been looking for the right person for some time. Would I agree to be interviewed for that position?

I did. And I was hired. I transferred to the Office part of the company.

Chapter 7

This American system of ours, call it
Americanism, call it capitalism, call it what
you will, gives each and every one of us a
great opportunity if we only seize it with
both hands and make the most of it.

—Al Capone, American
gangster

SAME COMPANY, DIFFERENT WORLD—OFFICE SYSTEMS

I'd been at Digital for nine years. The 1990s were beginning and, with no change in my career level or salary, I'd decided to step into something new: my biggest management challenge so far.
I was to be in charge of the "Office Systems Group"—250 engineers scattered around the world: 100 in Reading, England; 25 in Val Bonne, France, just above the French Riviera; the rest in Nashua, New Hampshire, in the building right next to where I'd been working since I'd arrived at DEC.

I would be responsible for all the company's office products: their word processor, spreadsheet products, calendars and clocks, the combination of which provided the basis for powerful business systems for our customers. The office group was already bringing in a lot of revenue from 13 products. And at the same time a fantastic team of researchers were developing and building a new office technology—some of the very first client/server technology—a software architecture in which the "client" system initiates a request that is processed by the "server" system within a computer network. (Think of your computer as the client and Google as the server.)

One of Digital's biggest products from the Office Systems Group (it had nine million users!) was a software offering called ALL-IN-1, an ingenious combination of capabilities that was the "Microsoft Office" of 1990. It contained a word processor, spreadsheet, calendar, mail system, document file system and a set of other "office" tools. (It actually provided a better mail system and word processor than I have today with my 2013 Microsoft products!) ALL-IN-1 customers loved that they could customize their business applications to fit with our products in one "set"—for example, imbedding the spreadsheet and word processor into their business applications—which would also eliminate having to retrain employees who already knew the basic ALL-IN-1. It was a super product for customers, but also for Digital, because it worked on a mainframe at that time. People connected their terminals to the mainframe, and since each system administrator's job was basically to keep all the mainframes working, there was virtually very little to maintain with the terminals. This made the overall cost of doing business quite economical for big companies.

THE JOB BEFORE THE JOB: "ACROSS THE POND"

My first challenge came even before I reported in for my new job. I was to start in January of 1990, but in December I got a call from Greg, my soon-to-be-boss, who said a decision was going to be made the week of December 20th which would affect what I would be managing for at least a year, and that I should be part of the decision. Could I fly to England on December 20th and fly back on the 24th? Let's see now: I am trying to close off my old job quickly so that I can move on, and my husband and children at home are counting on me to figure out how we can all have a good Christmas—and you want me to hop on a plane and cut out for *basically all of Christmas week?* Yes, that's right.

And I did. When I got to Reading, I found that a version of ALL-IN-1 was scheduled to ship in two months. Announcements were imminent, so the question was, "Should it ship?" Greg asked me to make that decision—and he wanted it made by January 1st. So, knowing no one, I walked into the 100-member Engineering group that was responsible for the ALL-IN-1, introduced myself as their new boss (but not quite yet) and started asking questions. I asked questions and listened for three days. We sloshed through all the meetings, I got my first, too-fleeting views of my engineers across the pond, and then I went home. Somehow, I arrived home just in time for Christmas Eve.

What I'd heard in Reading was that "the version about to be shipped did everything it was supposed to do" and "it had everything in it that was supposed to be in it." Completed as requested, as the specifications had stated. And everyone who answered those questions showed no enthusiasm whatsoever; I could get nothing from them other than that they had done exactly as they'd been told and it was finished.

But I wanted to know: Was it good enough? Would customers be happy when they got this version? Versions only came out once in a while and customers had been waiting for this one. Would it be worth their while to pay for it and to install it? Since I knew little about the details of the product, these were the only questions I could ask. But when I asked them to everyone in Engineering, I never got a "yes."

I believe Greg wanted me to go with him to England to look things over because I'd have to carry everything out, and he wanted me to really be involved in the output of that final "ship the product" meeting in January. But he certainly didn't want me to *stop* the product from shipping. In his mind, shipping anything was better than shipping nothing, because his experience with engineering told him that if for any reason we didn't ship the product, it would be a year or even two before he'd see product shipping again.

What the team had done was to exactly follow the rules, and they'd implemented a very difficult enhancement—one that was hard to do and should've made them all proud to have accomplished it. But I think they knew that few people would use the new capabilities. This version was to have been a "limited edition," and so the engineers didn't fix many of the bugs that had come in, or simple things that would have made a customer know that their product had been lovingly cared for. For example, as a customer I might have reported that on one screen there was a spelling error; the next time I got a new version, I'd sure look to confirm that that spelling error was corrected. But we'd done none of that. Only a few enhancements of very "hard stuff" had been worked on. Nine million customers were about to receive this new product, and when they got it there would be little they would see that was different. To me, that announced "this is an old product and we're not taking care of it."

I flew back over to England for a few more days the first week of January. I got to know my people and my product quickly. Then I returned to the U.S. and started meeting with our ALL-IN-1 Marketing Department. For the first time at Digital Equipment Corporation, I was part of a business unit and we were going to make a business decision; deciding to ship wasn't an engineering decision. The "big meeting" would be in a few days, so I worked with the marketing managers and shared my thoughts on the issue with each of them. I said my vote would be not to ship. They were appalled, but Rick, the senior manager assigned to me (he was my marketing arm), spent hours with me. I contended we should stall shipping of the product by six months—two months to fix all of these pesky bugs so people were proud of the product, and then four months of back-end testing, a reasonable amount required for this size product because it couldn't ship very often.

The morning of the big meeting, I first met with Rick and his boss, Ed, the head of Marketing. It was obvious that Ed wanted this product to ship right then, because he wanted something new to sell. I remained obstinate. I would do what was decided, but my recommendation was not to ship. I wanted a slip of only six months, I argued, and I would make that time worth their while. Finally, Ed said, "I want to ship now, but I don't vote by myself. I will support whatever Rick decides, because he's done the legwork on this one. He turned to Rick.

Rick, of course, was *very* uncomfortable. His boss had just told him how he wanted him to vote here. But Rick looked at me, and then said to Ed that he believed that Engineering could actually ship a much better product if they would delay it as I suggested. He felt customers would be much more satisfied, the winter months weren't big for new products anyway, and he had confidence in my ability to manage this so everyone *would* see a product this year. Rick said, "I vote to let Bev have her six months and ship the product the beginning of July." Ed just looked at the two of us, and said nothing for a moment. Then he grumbled, "Meeting is over," and walked out.

From there the three of us walked into our big meeting. People were on the phone from England and sales people were online from all over the world. My boss, Greg (who was Ed's boss), led the meeting. Greg asked me if the product was ready to be shipped. I said, "No." He cringed. I proposed my six-month slip and then reinforced my position by showing exactly what we wanted to do, and how we'd need the quick support of all the field to assess "what their customers saw as their biggest problems." We'd allow ourselves two months to solve the majority of those issues, then close the box and begin standard testing. The room quieted. People on the phones and online said nothing. Greg looked at Ed. "Do you support this foolish Engineering plan, Ed?" Shifting awkwardly in his seat, and without looking Greg in the eye, Ed said, "Marketing supports the Engineering plan. We'll make do until this product ships."

Greg glared at me with as much gall as anyone had ever done at Digital. He said, "If my people support your plan, then I'll have to go along with it. But let me say this. I have never seen the box opened up on this software product when it didn't take more than a year to close. So to me you are promising something that can't be delivered. But here's your one chance. If you can do what you've outlined, and can deliver this product in July, then you have my go-ahead."

And I had the gall to stand up and say, "Greg, I've committed to shipping this product in six months. *When* we ship in July, I'll expect you to toast the Engineering team with champagne!"

PROMISES, PROMISES

What followed was a harrowing six months. I flew again to Reading to meet with the entire ALL-IN-1 Engineering team and we prepared a plan to get all the input we needed from our best field people about what was aggravating customers most. We identified four really hard problems, which the engineers readily acknowledged. And we found "bugs"—more than 800 of them. Getting this product cleaned up without breaking things was going to be a real challenge. I put all of Engineering (close to 200 engineers) on hold for two months— *everyone* was going to work on this release. We dumped the bugs into buckets – hard, medium, easy and trivial—and distributed the hard and medium. I had to trust a few senior people to assign the best engineers to the hard bugs, but we needed at least one great engineer on each problem. On the corners of all of the cubicles we hung little baskets filled with the "trivial" bugs. When people were burned out from thinking about their hard problems, we challenged them to grab one or two of the trivial problems and get them out of the way. And because we had nine million customers, notices about more new "bugs" and requests for new features kept coming in. Lots of them. Continuously.

Every six weeks I spent a full week in England. I was constantly participating with large groups of people in worldwide meetings—with Marketing, with Sales. I was in person in the New Hampshire meetings, or phoning back there from England. People world-wide were on phones during these meetings, but it was a disadvantage for me one week in six to not see the key players in New Hampshire "eye to eye." I had to get used to talking to people I couldn't see.

We gave this plan our best shot. We fixed more than 800 bugs and problems that the field engineers identified, and closed the box on our work in exactly two months, when testing began. Engineers would then fix anything that overnight testing found, and we kept modifying our product and pushing ourselves to be faster and to do more. During the four months that we'd frozen our code and were testing, our industrious and extensive customer base submitted more than 200 new "problem reports" to the group. Having completed 800 fixes, suddenly a couple hundred new ones didn't seem like a lot any more. But we anguished: Would people like it anyway, if all those problems were there when it shipped? The 200 new gripes we'd received were quite obscure and we realized very few people would find corrections to them useful. We pressed on, beta-testing our new version in customer locations by the fourth month—and people seemed to love it. Whatever their favorite gripe was about this product, it had been fixed, and their opinion of the ALL-IN-1 soared! We were ready get this version off our hands, and get on to building better things for this product.

The Reading engineers once again felt proud of their product—a feeling that resulted from helping decide what creative efforts they could make to enhance the product and how good the product was going to be. They had not merely taken marching orders to someone else's specifications. Martin, the senior technical person in Reading, had been invaluable in monitoring every single thing that went into the product. I told him he was our "CTO"—our Chief Technical Officer. Martin and all his engineers were justifiably proud of the outcome. They were a hot team and had done a fantastic job.

We shipped in July. I was in England at the time and Greg made a point of apologizing to me that he couldn't come over and personally shake the hands of each engineer on this project. So I did it myself. And I recorded a tape of Greg congratulating the team, and brought it to Reading, and played it as we celebrated what we'd done!

CULTURE CONUNDRUM

Back at home, I'd been reviewing my experience working with Martin, this outstanding, creative engineer who could so effectively mold a product being created by a hundred people. I was so impressed with his work. But being a proper Englishman, he was quiet, doing his work behind-the-scenes, without fanfare. Martin was a "Principal Engineer," a fairly senior engineering position in the company that was achieved through promotion from one's boss—certainly a level from which to retire proudly, but, I thought, not fitting for this man. If Martin had been in the United States, for the work that he did he would have been at least one level higher: "Consulting Engineer," a high honor because so few engineers received that promotion. To be chosen a Consulting Engineer, one had to be recommended by one's boss and then the decision was made by a group of senior engineers in the company. I believed that over the years we hadn't promoted Martin appropriately, so I put him up to be considered for "Consulting Engineer."

The U.S. board that made the selections consisted of engineers who all worked in the U.S., and the group had a very specific set of stated and unstated criteria to determine who merited the Consulting Engineer title. Martin's whole personality didn't fit their picture. In the U.S., consulting engineers weren't quiet. They questioned boldly. They were more "in your face." They were assertive, almost aggressive. In the minds of the board members, those qualities were surely what made these particular engineers get things done. It seemed to be the way we "grew" really strong technical people in the U.S. But someone in England with that kind of personality just wouldn't get anywhere. (At one point when a fireball corporate Consulting Engineer had joined us at a meeting in England, I saw that he wasn't able to make any headway at all with my group of engineers there.) I don't think the U.S. board had ever received input from a manager before, but I went before them and, in my own assertive way, pushed them to consider "culture" along with their other criteria. They *had* to understand where those engineer candidates from other countries were "coming from"—and what made them effective. Taking those issues into consideration made sense to me.

Several years later, having given all of my information to my successor with an urging to push Martin's promotion, the board gathered again, and I got a call from my successor when they announced their results. Martin did get his promotion. Not in a timely fashion in my opinion, but he did eventually make it.

WORKING FOR MARKETING, AS ENGINEERS SHOULD

Managing the Office group was a different kind of a job for me at Digital; I was now working for the manager of an entire business unit. Smart and savvy, Greg knew about business. He'd built ALL-IN-1 into a product that served and satisfied nine million users. He focused his discerning insight on making sure we were always building and maintaining the right products and doing what was "smart business." I'd never had someone with this kind of business acumen guiding me at Digital Equipment Corporation, and I couldn't learn fast enough!

Greg gave me a finance man who told us each month how much of our product was sold, country by country, and Greg advised me how to respond accordingly to those numbers. If a product wasn't selling enough, I should reduce the number of engineers working on it, or add engineers to give it a feature that would sell more, or market it differently. Everything had to make business sense. I lapped it up, and had to learn how to deal with justifying everything I was doing: Did it make marketing sense? Business sense? I'd never had to fight for what I needed and defend my reasoning for what I wanted so thoroughly before. In meetings with my peers, I was the one Engineering manager competing with six Marketing managers—and I made sure I always came prepared.

While we had products shipping and the world was working on mainframes, personal computers (PCs) were coming into their own. I spent a great deal of time working with key engineers on my teams to develop a full client/server technology so that existing customers could keep Digital's mainframes to use as servers that kept their important data, and then attach PC and Mac terminals to their mainframe. This was started before I joined the group, and I could see just where this could take us, so we left our best engineers doing this key development.

We started other projects to grow an office product we called Teamlinks, which would also work on PCs and Macintosh computers, connecting to the mainframe as their server with our client/server code making everything transparent to the customers. This was such leading edge technology—and just what industry needed at that time—that I was ecstatic. Our team delivered, and was ahead of the industry curve with our technology but, sadly, we were ahead of the rest of Digital Equipment Corporation as well. DEC didn't stand behind us on Teamlinks, and internal operations of the company made it hard to get marketing support for this fantastic product. Digital wasn't ready to promote these PC products even though our customers wanted them.

PEOPLE PERKS—KNOWN AND UNKNOWN

I'd gone from being scared out of my mind that I couldn't handle this job managing so many people—and the fact that they resided in three countries on two continents—to loving my job and the smart people I worked with. I was lucky to have inherited a secretary who was a delight and a godsend. She saved me often: what a joy to have someone keeping track of all I did and be on top of it! She'd been in the group much longer than I had, and steered me in the right directions, making it easier to be responsible for more than 200 engineers.

Little did I know I'd also been lucky in another way. Years later I was offered a General Manager position in a Fortune 500 company in the U.S. I learned then that my resume had come across the desk of one of that company's General Managers, an alert fellow with a good memory who, in 1990, had been managing a group of engineers for DEC in the same building in Reading, England, where I'd overseen the ALL-IN-1 rescue and recovery. He had observed what I'd accomplished with my British engineers, and over the many months of that experience I touched base with him whenever I returned to work with the group. I never would have imagined that other significant opportunities would come to me years later from someone who I'd never worked with, but who had only watched me and nevertheless evaluated my work.

GATHERING STORMS

At this time, when my Office group was thriving, turmoil within Digital Equipment Corporation caused a number of factions to arise. Evidently Greg was not in favor among the senior executives. The problem was that no other Engineering groups were driven by Marketing—throughout the company it was Engineering who led and built products, whether it was good "business" or not. Greg was by far the best marketer I'd ever encountered. He was a joy to work for because he was so focused on achieving success for the company and accomplished that by motivating and driving his group—Marketing, Finance, Engineering—toward our own market success. But the powers that be didn't like that. They wanted all of Engineering under one roof, and *not led* by Marketing. A political power struggle. So I awoke one morning to find that Greg had had to cede all his engineers to a central software engineering group, made up of all groups representing operating systems, compilers, databases and every application the company owned.

In this reorganization, to create one "big" Office group for the company, my Office Systems Group was blended with another, similar one consisting of about 125 engineers headed by Ian. That group also worked on "kinds of office products," but they had few products shipping, and, as a group, were not yet profitable. My group, on the other hand, was very profitable. When the dust of the group merger settled, however, I found that I was reporting to Ian, who had hired someone to replace himself as leader of his small, original group so that he could head our new combined organization.

I found Ian to be a good friend and a very nice man; I loved to talk to him about technology in general. He would have made a good drinking buddy if I'd needed one. But good buddy qualities are not always good boss qualities. Professionally and personally, I was used to being given specific goals that I had to achieve to drive business for each one of my products, and was accustomed to being managed continually, carefully and forcefully. Now I had a boss who never set a goal for me. Ian had no experience looking at the "business" of engineering and had no interest in how we'd made decisions to evaluate our products.

Furthermore, he had no experience with any of the products I'd been building, but nonetheless treated them as though they were the company's "old" office products. He, of course, had the "new" products—a WYSIWYG (What-You-See-Is-What-You-Get) editor based on a product we'd purchased, among other things—none of which had been profitable up to that point. Yet, with our two groups combined Ian still *looked* profitable, because my group was so financially successful that it carried them both. In the shifting milieu of that merged group, I had to work hard to protect my engineers, my business decisions, my products.

It was a bummer of a time.

A phenomenal Research and Development group was part of my team and at one point we were working with universities to capture the latest in technologies. I had products that ran on Apple and Windows "clients" connected to big DEC servers. This was perfect for the PC-driven, client/server world that was coming. Even so, Ian and I never discussed the power of these new products and how we could use them to catapult the company into the industry's growing PC sphere.

But to be fair, the problem wasn't limited only to Ian. Many people in charge just didn't understand business or business models, and as a result I had to continually justify the existence of *all* of my teams and products. I wasn't always successful. On the personnel level, I eventually lost my two financial people, because, it was reasoned, "Who in Engineering would need that?" And in the product area, my first "client" got stopped just as it was about to ship. It sat for three months while somebody (or a bunch of somebodies—I couldn't begin to tell you "who") in the company decided whether the product should really be shipped. It was very depressing to have really good products ready to go and our own management thwarting customers who wanted them.

There are technical companies that are market driven and those that are engineering driven. I'd been given a taste of the power and excitement generated by being responsive to the needs of customers. Now I was back to being engineering driven, and as an engineer I should have loved it. Instead, I couldn't believe how inept we were being about so many things!

At that time I was called before the Strategic Task Force (STF), a team that evaluated new products to determine whether they should be built. I'd been asked to present a new product that had already been in development in my group at the time of the Office Group reorganization. During ninety percent of my time before the STF, I received no questions about the product I was presenting. Instead, they focused disapprovingly on the products Ian and his group had developed, asking me questions like "Do you use your boss's WYSIWYG editor?" and "Did you use it to create this presentation?" At first I was confused. But slowly I began to realize that my boss was "on the carpet" here, and they were giving me a hard time as his stand-in. Clearly, the STF didn't really want to understand what I was offering: why client/server technology was so important and why PCs were where the world was going. The meeting had nothing to do with my product; they paid no attention to it.

Had they listened, they would have reasoned that this new client/server technology would allow all of their existing products to be part of what today is called the "cloud." Digital could have continued to offer the solution to keeping business investments safe back in the cloud while company employees were working on these clients. Storing customers' work in a safe place was how Digital had always excelled. Here was the chance for DEC to keep all its PC-using customers and move them impressively into the next phase of handling computers. The STF gave up that chance for the company.

Even more important for me in that STF experience, I realized that during eleven years at Digital I'd never before been required to come before the STF for any product that I'd built. Prior to this, the STF had never looked at *any* software products, so they had no basis for understanding; their evaluations for these products were meaningless. I left the meeting disillusioned: for all my years with the company and all the products I'd built, the STF had never cared to check on any of my completed, successful software products. I felt like I'd been to a Laurel and Hardy show and had wasted my time. It was disheartening to see the company being run this way.

Outwardly, Digital Equipment Corporation's growing reputation for its high quality software and operating systems meant companies around the world were betting their businesses on us, and were pleased. Internally, true to our history, we were focused on building new hardware and had teams ready to build it. Those hardware groups wanted three operating systems to run on their new hardware—the plan being that any applications built for any of those operating systems would be able to run on the hardware. It was a business model the opposite of Microsoft and other companies, which were touting themselves as software companies. Digital had such robust software that it was fantastic, but the corporate conventional wisdom saw those products as only a means to sell new computers. Even given all that, our software teams were so capable and strong that I thought we could weather the storms.

TIP #7: CHANGE WITH CHANGE

Change happens. Go with it, and don't let it get you uptight. You can't keep things the same, and you don't really want to. New challenges are workable.

Chapter 8

*Whatever your hand finds to do, do it with
all your might.*

—Ecclesiastes 9:10
*New International Version
(NIV)*

OPERATING SYSTEMS

For the next couple years, the Office group provided me exciting challenges and smart, engaged people to work with. But wild winds of change continued to transform the computer industry, and Digital Equipment Corporation was trying hard to keep up.

Digital boasted 15 million satisfied VAX/VMS users who purchased our hardware, used our operating system, and bought all the compilers, tools and applications we sold on these systems. The company never kept track of the separate and distinct worth of an operating system, as it was bundled with the hardware, but the hardware kept selling. The VAX/VMS systems had proven their worth. The VAX had a 32-bit "complex instruction set computer" (CISC) architecture, which worked by allowing a single instruction to simultaneously execute several low-level operations or do multi-step operations. This had been the power of a VAX.

But a new hardware system designed to dramatically increase a computer's speed had been in development since the mid-1980s and seemed to be more commercially viable. Digital wanted to be at the leading edge. It would prove to be a costly effort.

By 1992, with this new and significant shift in design strategy in the works, DEC was building a follow-on machine to the VAX. The new machine, initially called "Alpha," would have a "reduced instruction set computer" (RISC) architecture, based on the concept that simple instructions (instead of complex ones) allowed the computer to perform each instruction much faster. The power of the Alpha systems would be lightning-fast execution.

Construction had already begun on a multi-million dollar facility to fabricate Digital's new RISC chips to go into the Alpha computers. Even more exciting, these new chips would use a 64-bit architecture—a third instance during my lifetime of doubling the power of computers! In addition, the ambitious hardware teams planned to build not one initial RISC machine, but *at least five*—of different sizes—at the same time.

Digital had become accustomed to producing a new VAX version and voilà!—all the software developed in the company and all the customers' applications would run on the new version because the VMS Operating System (optimized to run on a VAX CISC system) ran across all DEC's hardware products. But these new RISC machines were different. Building VMS on hardware other than a VAX, much less on hardware that was so fast primarily because it did *none* of the things that a VAX did, greatly magnified the task for engineers.

And from the point of view of existing customers, if they were going to buy these new RISC machines, they would naturally expect all their CISC applications to run on them. The challenge for us in software, then, was to create a new "Alpha VMS Operating System" that acted like a VAX/VMS Operating System so Digital-built applications as well as customers' applications didn't know the difference. Digital also wanted their UNIX Operating System and Microsoft's NT Operating Systems to run on these RISC machines, but VMS had to come first.

Without VMS (and the millions of applications then in use) being able to successfully run on these new "beasts" called Alpha systems, the hardware would not sell. The huge customer base would abandon the company and Digital Equipment Corporation would go out of business.

That seemed so obvious to me.

About this time, several events occurred that would have an unforeseen but long lasting, adverse impact on the company, and would turn out to affect my career at Digital as well.

The overall leader of all Digital's software—a powerful man who'd come from Europe to take that position, and who'd likely had terrible software/hardware battles at high levels—left the company. Within six months he died of cancer. Also, one of the key software proponents at all of the big Alpha meetings also died of cancer. When two of the most senior managers in software end up leaving the company either in frustration or by *dying*, it was obvious that it just wasn't easy to be a software proponent at the highest levels of Digital. Now senior software positions were filled with the managers from the hardware parts of the company. There were no "software people" left in senior management.

It is one thing for a company to have to deal with the unanticipated departure of critical employees. But it's quite another for the company's executives to usher prime employees out the door. That's what happened next.

Digital decided to offer a very lucrative "package" to everyone over 50, regardless of the importance or value of their work to the company. The offer was quite an incentive, and the manager of the Alpha VMS OS group was no fool. She took the package and left the company.

In a quandary, suddenly the Alpha VMS team of 150 and the many groups tied to the work they were doing had no manager. Also, the remaining engineering team wanted to hand-pick their manager. A senior group of these engineers had gone to their boss's boss and made their case that without the right manager the team definitely would not make the impossible deadlines set up for them. This senior group had a lot of influence in the choice—a fact I wasn't told at the time.

When I learned of this management opportunity, I knew I should apply for it—this was one of the times I felt compelled to pursue a job within the company. I thought I could make a difference and I didn't think Digital could survive if this project failed. I loved my Office group, but I could see the risk in RISC for Digital. I sought the interview and was chosen for the job.

My new boss was Chuck, a hardware man who had come over to head the VAX/VMS unit and thus also headed the new RISC group, called Alpha/VMS. Reporting to Chuck was kind of strange. He was obviously uncomfortable with me and had a track record of not working well with women, so I should have surmised that, at least from that perspective, this would be a strange situation to step into. Once I was walking down the hall, spotted him standing there and said "Hi, Chuck"—and he didn't even answer me. But a split second later, he said "Hey!" to the man walking directly behind me and start talking to that fellow about anything and everything. In a meeting situation, however, Chuck would call on me like he called on his other reports. At least there I was seen as an equal to my peers.

I reported to my boss on our group's progress and attended his staff meetings. Chuck stayed out of my way and just let me work with my team and all the other teams in the company who were adapting their applications, tools and compilers to this new operating system we were building. Essentially, he didn't bother me, but neither did he provide any guidance. Chuck knew so little about software that he didn't even attempt to give me any direction. But I couldn't spend energy worrying about my relationship with my boss because I needed to jump head first into the deep water. The job the group had bit off was so large and difficult that I barely had time to register the disconnection—and time was critical. The Alpha chips were already in fabrication and we needed to have an Alpha/VMS Operating System tested and ready to ship in about six months.

(Quite some time later, while stuck in the Newark airport in an ice storm, I ran into an old friend, Corey, who greeted me with "Hey Bev, haven't seen you for years! I heard that you were the kook who agreed to manage the building of the Alpha VMS Operating System. No one in their right *mind* should have taken that job—you know it!" When I responded, "But we did it," he replied, "That's no excuse! You were *stupid* to take it on!" And I'm sure Corey was right.)

It was June; Version 1 of the product needed to ship by November. Version 1.1 would follow six months later, to include only minor fixes that we'd find once people were using Version 1. Any new functionality would come two years later when Version 2 was planned to ship.

Digital had done a lot of risky things on this project, like constructing our own "fab house"—an immense "clean" building to be strictly used for creating our own Alpha chips. This effort was unbelievably expensive and would only be offset if we sold millions and millions and millions of systems. But I assumed that the company had sound business plans, and instead focused on what I needed to accomplish: creating a new VMS that was identical to the VAX/VMS, which from day one would run on the AlphaVMS system. Fifteen million users would be depending on that. I had to take one step at a time.

It had been decided that we needed to build simpler, but still VAX-y, hardware structures around the RISC architecture to allow RISC to remain fast but give it more VAX-like capability. Most of these structures, which we called "jackets" and "vests," were complete when I took over, but early tests showed our final product, with all these added components, made our AlphaVMS even *slower* than the VAX/VMS product. So customers would buy a fast RISC system that really ran slower than the old CISC systems? I didn't think so! We rolled up our sleeves.

HOW AND WHEN YOU CROSS THE FINISH LINE DEPENDS ON THE TEAM YOU HAVE

I chose Moxie, a Corporate Consulting Engineer (Digital's highest level engineer) who was no-nonsense and a careful, thorough, analytic thinker, to head a performance team that would closely examine the operating system development. The team would improve performance as they proceeded, and provide hints and options for engineers in the groups to use as they worked, to find more ways to make programs faster. This team stimulated hundreds of other engineers to optimize for performance. All the operating systems teams as well as the applications teams pruned and tightened and firmed up their code. If we were shipping in six months, then we should have been pretty well finished and testing and waiting for the final prototypes from hardware to do final testing. But we weren't.

Meantime, I was working against the clock with every team to finish the work they'd begun; we held daily 15-minute meetings each morning to share whatever "broke" in the nightly build of the operating system. We included in these meetings a developer who represented each application that was bundled into the operating system, like the debugger and other products I knew well. These applications were using very early versions of this operating system, and they needed to be updated by fixes we were making daily. Eventually, the engineers who built the compilers and the key applications joined the 150 engineers who worked directly for me so that each morning we all would get the same essential information.

I found that my earlier years spent at SRL building software so that it would make the most of hardware held me in good stead! This new the low level operating system needed to take advantage of everything these new RISC machines offered. If there was any feature the hardware people were building into the chip that we didn't capitalize on, then we might not have a strong enough operating system. We built drivers that called for every capability of the hardware. We kept producing jackets and vests around what little we had so that the new

Alpha/VMS was as rich and effective as the previous VAX/VMS systems. Remember the story of the emperor who had no clothes? Well, this RISC architecture was a very naked one! And unless we clothed it elegantly, it was going to be one pathetic system!

During this crazy six months it felt like I was keeping a hundred balls in the air all at once. More than 300 engineers involved in the building of related component parts were all tied in by phone or with representatives to all the meetings of our Alpha/VMS team. Our performance team was running all over, meeting with all these groups to optimize for performance. Performance engineers worked with any application having problems…and while they were busy working with those other teams, they were losing time on their own software development tasks.

The hardware group had by this time planned to have *five* different hardware machines all shipping on the November date. When chip development started, the hardware plan had been to put the Alpha chip into only one new computer. That seemed enough of a task at the time. But as development continued, other hardware development groups also wanted to create their own size and shape of a RISC computer to run with the new chip. By the time we were nearing our deadline there were five different-sized computers that would all ship as soon as the Alpha chip was fabricated and inserted into each of them, and as soon as there was an operating system available to ship with the chip. This meant that the new Alpha VMS operating system—in all its aspects and including everything built on it—had to be fully tested on *each* of these five computers, a fact that increased my small quality-assurance team's workload far more than fivefold.

The leader of the QA testing group, Ira, had his work cut out for him. We didn't have expensive simulators, and we didn't have the bodies to man those machines if we bought some. Ira had to figure out how to do partial testing, test inventively, and then be ready to *really* attack the machines once prototypes arrived, so we would have enough hours remaining to finish basic testing. We had some testing done in Israel and some in Ireland—and everywhere people found creative ways to test the product effectively even before it was complete and ready to be tested.

Then along came November and the dust settled.

Version 1 was a wonder! Our Alpha VMS invention (renamed OpenVMS/AXP) worked, it worked well, and it looked and felt like VAX/VMS—close enough that users couldn't tell the difference, and better yet, neither could their applications. We shipped the product working on all five RISC hardware systems, as promised. Truly an incredible achievement.

But delivery on time, as scheduled, of something so very difficult was regarded by Digital Equipment Corporation as merely a matter of course. As I shipped the result of what I felt had been an extraordinary effort, I was asked to immediately lay off 10 percent of my team.

Wait. Shouldn't this accomplishment involve a bonus? A rousing cheer? At the least a handshake? For me, managing people wasn't fun when I could justifiably pat my people on the back for a job well done, but then my company reciprocally didn't seem to have my back. Initial sales had fallen short of the business plan expectations, but solving that issue by cutting 10 percent of our groups across the board seemed like bad management. The teams had just performed miracles. Uncomfortable and unpleasant as it was, however, I cut the team as requested, and everyone moved on.

Before Version 1 had shipped, I'd had a number of talks with our insightful product manager, Bob, who realized that once these OpenVMS/AXP systems started shipping, people would want to take one and put it in their VAX "cluster," the name we'd given to an arrangement of VAX machines that had been purposely connected together to operate as one system. The beauty of the cluster concept was in getting maximal efficiency from the hardware. When two or three—or ten—VAX/VMS machines were connected together, they would act as one big machine with interchangeable resources, moving jobs to the most available machine, for example, to achieve greater productivity.

Clustering the new Alpha system in any configuration hadn't been anywhere in the initial, official Version 1 tech or business plans. We'd often discussed a plan to eventually develop those early Alpha/VMS machines into clusters to be delivered in Version 2. But that was scheduled to happen about two years after the Version 1 ship—and on a collection of *only* OpenVMS/AXP systems, an issue that, by itself, would be hard enough to resolve. Creating clusters among *mixed* VAX and OpenVMS/AXP systems, as Bob was proposing, would be an even more formidable task. But we knew that would remain the ultimate goal. Though the cluster group was already hard at work on another difficult problem, unless they could make OpenVMS/AXP systems fit into clusters of VAXes, our engineering group didn't see the new systems selling well in the long run.

The issues remained after the Version 1 release and we knew we would need to revisit them quickly.

In our engineering meetings we talked about the cluster concerns and considered approaching the problems from opposite perspectives. Should we try to first create a cluster of OpenVMS/AXP systems, then add one VAX/VMS machine to it? Or should we begin with an existing cluster of VAX/VMS machines and add just one OpenVMS/AXP machine to that cluster? Optimistically, and perhaps naively, we wanted to deliver both kinds of these "mixed clusters" as we called them, even though neither proposal had yet to appear anywhere in our business plans. Mixed clusters just didn't seem to be feasible. It was like trying to physically connect Lincoln Logs and Legos and expect to be able to build anything.

In one meeting Ralph, a quiet engineer, started to articulate an idea he'd had, when he was drowned out by the others and his thought was gone. But not before I'd heard it. I asked that I meet with Ralph and with Gordon, the clusters manager. We talked through Ralph's idea, and it had some merit.

I evaluated the speed and smarts of the engineers regarding what would be needed to deliver clusters, and we decided to give Ralph's imaginative notion a chance. We still didn't know that it could work, but we had no other plan that could produce clusters within two years, and do it without an inordinate amount of work in that short time.

The whole team grabbed hold of Ralph's concept and, enthusiastically applying a lot of careful thought and diligent effort, they developed and embellished the design. The quality team, too, had worked against incredible odds to test these clustered systems. That was true particularly in just getting enough systems to do thorough tests, as new systems were just becoming available. We were often lucky to get one or two prototypes of a new machine, even though to test clusters properly, each testing engineer really needed a slew of machines.

Finally, after solving the many remaining issues, we shipped OpenVMS/AXP V1.5 *with* VAX/VMS *mixed cluster support* a mere half-year after we'd shipped Version 1! What a team to work with— they'd powered up and in six months accomplished more than they initially believed would take two years to do. Phenomenal! Customers could buy a RISC system or two, test critical business applications to become confident of the results, and then embed V1.5 into their VAX clusters without losing any of their initial investment. Our clusters implementation worked. The software engineering team was justifiably proud of this feat and, I felt, deserved a great deal of credit.

But once again, my Digital management disappointed me. The corporate "ladder" above me was by then bedecked only with former managers from the various hardware groups. No one in upper level management in all of Digital Equipment Corporation knew software or truly held software interests, experience or advocacy in mind.

As our remarkable clusters version started shipping, my manager required me to cut my engineers again, by another percentage. No recognition for the team, no bonuses, no hoopla. The company needed to reduce its expenses; computers were getting smaller and cheaper, and profits were down. How better to solve that problem than by letting go of some software engineers? I had a group performing miracles, and no one above me could recognize the value of what the company was getting! Because hardware managers were in charge of all of Digital's software, they really couldn't understand the achievements in software that we had accomplished for our customers and for our company.

GLIMPSE OF THE FUTURE; NOSE TO THE GRINDSTONE

A couple things happened during this time that made me think about opportunities outside Digital Equipment Corporation, even though the company had provided me fantastic jobs for 13 years. I was tapped for a job interview at a high-powered national networking company. Unsolicited. Whoever had recommended me, I was part of a group of about eight candidates. I decided I'd go, mostly because I hadn't interviewed outside of DEC for so long; I thought I needed the experience.

In fact, I had always told my developers to interview every couple of years. That way they knew they wanted to stay in my group! I was cocky in believing that they had to want to work with my teams in order to give their all. I figured that by accepting the interview I was taking my own advice.

It didn't trouble me that I wasn't a networking expert; I knew software management. I went through the first round of interviews not certain I even I wanted the job. The callback came. They were down to two choices, and I was one of them. More interviews. They finally chose the other candidate, a network expert from their competitor, Cisco. I realized then that although my experience might not have been a match for *this* job, there were interesting jobs out there that would be well suited to my skills and that I was very marketable. After 13 years at Digital without ever thinking of leaving, the thought had been planted in my mind. But even then, I didn't put my resume out.

I also lost one of my best engineering project managers to Microsoft. He came to me, showed me the data on sales of our RISC chip, and calculated how many of these chips would have to be sold for DEC to be profitable in this endeavor. He concluded that Digital didn't have a chance of pulling out of this one; the company had a flawed business plan. I listened, and his facts were very accurate. This engineer had small children and needed solid, reliable employment. I was sad to see him go, but I understood. Yet I still wasn't ready to look for a more stable company. I loved the smart and savvy engineers in my OpenVMS/AXP group and we had hard work to do that I knew I could make happen.

By March of 1993, the OSF/1 UNIX Operating System for Alpha systems shipped and six months later Windows NT Operating System was shipped for Alpha systems. Thus, by September of 1993 Digital finally was selling what the company had originally planned to offer the previous November: all three operating systems working on the new RISC architecture. Thank goodness my engineers had been able to give them one of these three operating systems when the hardware had been ready. Meanwhile, we kept shipping V1.5 with clusters and began work on V2.0, a release of fixes intended to make the new OpenVMS/AXP architecture virtually indistinguishable from VAX/VMS. The issues in this release were complex and always interesting, and I think we made excellent tradeoffs. It was very hard work. In the end, customers got identical capabilities for both the OpenVMS/AXP RISC product and the older VAX CISC product, even though they were on very different underlying architectures. Big achievement. In May, 1994, we shipped our Version 2 product as OpenVMS/AXP V6.1 simultaneously when the VAX/VMS group shipped OpenVMS/VAX V6.1. That presentation showed them as identical.

OpenVMS/AXP V6.1 was very well received. But it was not the only focus. Digital senior management was pushing the other operating systems, which now ran on their new RISC architecture—in general, was not a bad idea. I didn't realize it, but in the process of assessing operating systems, my boss was not aligning himself with VMS and with what we'd created. Chuck didn't understand the power of VMS. He thought that other "newer" Operating Systems, like UNIX, deserved more attention and neglected to support us to senior managers. He never went to bat for us. That would become apparent very soon.

BIG-HARDWARE-COMPANY BLUES

Meanwhile, on the wave of highs that accompanied the successes of my team, I had three other strategic ideas I felt we could try that would profoundly benefit DEC and our customers. Chuck gave me no encouragement here. But even if he couldn't imagine progress, I could.

Once we had freed our VMS Operating System from VAXes and it could run on RISC systems, I knew we could run it on anyone's computers. At that time Digital's hardware systems could run VMS, UNIX and Windows NT operating systems with any software applications. My idea was to take OpenVMS/AXP—which had the most users—and make it run on whatever hardware the customer chose. That was the Microsoft model, and I loved it. Now DEC could do that. In fact, we'd already prototyped running OpenVMS/AXP on Intel systems. Need secure systems for the world? Digital Equipment Corporation would now have them!

But my boss refused to back me. No explanation as to why we couldn't. No going before the STF because it was strategic to the company. No discussion. Just "No." So I went back to building VMS systems only for Digital computers.

The second strategic idea that had such potential was to focus on the needs of application programmers: how could they thrive using these

RISC machines? I assessed the OS teams—NT in Seattle, UNIX and VMS—and was able to pull together a small team, taking two experts from each group, plus myself. We seven sequestered ourselves for a week, as I'd negotiated with my manager to disappear for only that length of time to work on this. The group determined the feasibility of building a "single jacket" for all three operating systems. This would allow applications developers to write *to the jacket* and the work would automatically run on all three systems. We proved to ourselves that this could be done and were *so* excited with our work. I stayed up all weekend writing up the proposal so that Chuck could present it at some big company meeting the following Monday. Unknown to me, he never presented it, but neither did he tell us that. Had our proposition been accepted, and had we been allowed to build it, Digital Equipment Corporation would have had a powerful way to sell their hardware. And their software. But that was also a "No."

So, on to the next idea. I was ready to start exploring and using the new power of a 64-bit architecture. OpenVMS/AXP mimicked all that VAX/VMS had, so now that we'd ported to RISC machines with a 64-bit architecture available, why wouldn't we use it? Digital could *explode* the real values of RISC. We'd be the first company to do that. We'd have power to blow away what any other operating system—including our own—had ever done in the industry! Yet initially Digital wasn't interested in our doing anything with VMS; they didn't understand its potential.

But I understood, and was excited to begin to drive this new technology. I'd been working with our kernel group and its project leader, Lyle, to modify the kernel from the 32-bit VAX system to take full advantage of RISC architecture. For Lyle and the rest of the kernel team, this would be a really hard assignment— probably the hardest of his career, Lyle understood—and they all went right to work. By the time it was conquered, I was no longer with the company, but Lyle contacted me to tell me that his team had met the challenge. Software that took full advantage of 64 bits in the kernel was complete. It worked well, was really powerful, and was ready to be shared at a users' conference and shipped to customers. I was so proud that we'd started this project, and that the team had been able to pull it off!

Chuck decided about this time to abandon VMS, accepting a job to head the UNIX Operating System group. But before he left, he merged the VAX/VMS and OpenVMS/AXP groups into one, considering that VAX/VMS and OpenVMS/AXP were now identical in what they did. He felt that the teams could function under one manager. In almost his final act on behalf of this formidable, complex and highly achieving operating system, my boss named the junior fellow who headed VAX/VMS as the leader of the new group, thus eliminating my job.

The engineers whom I'd encouraged and worked shoulder-to-shoulder with, and who had given their every minute to this project for so long were now going to be handed over to someone else. Not my responsibility anymore. The huge success that we had in pulling off all that we had was muted by the fact that I was finished working with this group. In the world of politics within a company, I had lost. In trying too hard to achieve *my* purpose—getting the projects completed well and on time—I hadn't played the game of company politics. Chuck was more comfortable with those whom he knew well: hardware people. We software people made him uncomfortable, and I knew I had done little to change that.

Each time I'd moved on at Digital, it was for "the company's good," not because it was a great promotion for me, or even an increase in my salary. I didn't know how to play the game well enough. But, boy, had I had phenomenal technical challenges, and the power to meet them with high-powered people at my disposal. I was given grand opportunities, and I'd taken advantage of them. They just weren't good for advancing my career. If I had to choose to accept each of these jobs again, I'm sure I'd make the same choices—I'd been attracted to the challenges and the fun.

I was too visible a manager for Chuck to let go, so he moved me to a non-engineering group to work directly for Al, part of DEC's top

Senior Management. I think Chuck thought it was a good move for me. Al was responsible for all software engineering for the entire company. This was supposed to look like a promotion to me, but it deprived me of my chance to work with the wonderful engineers I'd been with, busting our butts together to produce product. It pulled me away from that excitement and from the challenge of producing product.

Al was a hardware guy from way back, having managed hardware groups for years. I liked Al, but I recalled that a couple of years earlier he'd asked me some questions about using email that showed me he didn't have the slightest idea of what any software was or what it could do. But at some point Digital management, in their wisdom, had made Al the ultimate head of software! Al didn't understand the software for which he was now responsible, and to make matters worse, he was given 60% of the previous year's budget to manage the current software operations. He didn't quite know what he had or what to do with it, but he had to do *something.* His immediate plan was to just sell off as much of the software as he could to other companies—after all, Digital was a hardware company and we didn't need all of this software. He was in the process of doing that when I joined the team. My job was to help Al determine how to survive with 60% of his budget.

Now I was an individual contributor working this out for him, and without 200 people to manage, I was like a fish out of water. For the first time in many years I had 40-hour workweeks centered, disappointingly, on work that was boring for me! I took the extra time in my week to enjoy my family, and to balance my life a bit. It was a refreshing shift of focus.

BUCKETS OF SOFTWARE

Al wanted me to remain at his side to work on budgets, but said I would rise highest in my career at DEC if I managed companies who were buying our software, keeping track of how they were doing, etc.

He, of course, wanted this software to continue to work on Digital hardware. Al felt that he could save so much money for the company and reduce headcount if Digital just kept selling off as much of its software as possible, and then used people like me to ensure that good software still ran on our hardware. I suggested to Al that if he wanted people like me to stay in his company, he had to give me difficult software projects with good engineers to carry them out. I didn't watch other companies work.

That conversation made me recall one time when I'd been asked to get engineers to complete specifications for a product, but then was required to have the work done by another company before bringing it back to my engineers to test it to be sure it was right. They *hated* that! Management had given away the fun part, and left the engineers with the drudgery. That's the way *I* was now feeling about management: I wanted to be working directly with the engineers who were making the design decisions and figuring out how to best build the products. Why would I want to be evaluating products *others* had built to make sure they worked on our hardware?

Over a couple months I did a huge amount of work for Al in order to present him with a solution to his software dilemma that could be achieved with less than two-thirds of his usual resources. I processed my budgetary task efficiently, dividing all the company's software into three buckets according to their value: what was past use, presently profitable or held great potential for the future. I compiled lists and classified products. I researched, analyzed and rated personnel.

While I was doing all this, I sent out my first resume in thirteen years.

Finally, I sat down with Al and showed him all the specific assessments I'd done to categorize products, along with their descriptions and reasons why the products should be handled as detailed. I described what I'd evaluated and told him how to go forward with his "three buckets."

The first bucket contained software products out of date or past their usefulness to drive the company's business. These could be sold off and, if done quickly, the manager of this, the smallest, group could maximize revenue in a short time. This manager needed to know sales. I showed Al an itemized list of managers within the company who would fill the bill.

The second bucket held the software products Digital owned that were currently profitable and would bring money into the corporation for a good while. These were the wonderful "cash cows," products that our customers depended on for their businesses and which would continue to sell without major enhancements. I explained how Al would have to choose a manager for this group who would be rated on keeping customers happy with these products, with minimal dollars invested. This person would be able to find a few "cool" enhancements, keep the products looking great and serving their purpose, and could do that without much money for development. I gave Al names of managers and engineers I knew who would be good at optimizing income by engaging the fewest engineers, keeping costs down and revenue up.

The third bucket included all the company's software products that carried the most potential and were right on target for sale into the marketplace—the inventive software that required a significant investment of time and energy. Engineers would love to be in this group; it would require known innovators. So it would also require an entrepreneurial manager capable of growing the businesses of these hotshot products, and Digital still had a few of those people who could pull that off.

Al's eyes just glazed over. He was overwhelmed with this software business. With his background in hardware, he really didn't see why we were putting money into these software products. More to the point, he really didn't want to build the three groups. I would have been happy if he'd at least created the third group—that was the one I would have wanted to manage and I knew we still had other good managers who would have thrived making competitive products in such a group. But Al had made up his mind. And my resume was circulating.

He continued with the "sell off," but not in a smart way. He didn't understand the value of the software he had, and got pennies on the dollar selling software from the wrong buckets. Without many of those products his hardware was no longer viable. He never realized that.

So with several other job offers having come to me, and after 13 wonderful years building great software for a big hardware company, I handed in my resignation and left Digital Equipment Corporation.

TIP #8: RESPECT YOUR MANAGER.
No matter how great the job, you must respect your own manager in how s/he thinks and works. Respect what your manager values. If you cannot, you need to find yourself another manager.

Chapter 9

A human being should be able to change a diaper, plan an invasion, butcher a hog, conn a ship, design a building, write a sonnet, balance accounts, build a wall, set a bone, comfort the dying, take orders, give orders, cooperate, act alone, solve equations analyze a new problem, pitch manure, program a computer, cook a tasty meal, fight efficiently, die gallantly. Specialization is for insects

—Robert Heinlein, dean
of science fiction writers

FINDING THE BEST FIT

Building operating systems for DEC meant that I'd been talking to hardware engineers regularly, making our operating system take advantage of the hardware. I had grown so comfortable with both hardware and software that I could now tell a joke like:

"How many computer programmers does it take to change a light bulb?"
"Are you kidding? That's a hardware problem."

For some time I had had the feeling that Digital Equipment Corporation was "dying." Even so, I had needed to be kicked before I sought out other opportunities. That kick had finally come when my teams were taken from me and software concerns were consistently relegated to second-class-citizen status.

Years of developing camaraderie with fantastic engineers meant it was hard to leave Digital. I'd spent 13 years giving my soul to the company with each project and with each group I'd managed. Leaving the people would be very difficult. But I began to wonder about having to leave the work. I had always liked the fact that hardware created actual things, and I enjoyed building software to program the hardware to take actions. When the time came, I thought I should look for a job where both aspects were in the picture. It was 1994.

Reviewing the offers I received, I finally settled on two that were excellent. One would give me a good salary and, if I stayed in the position for three years, a $300,000 bonus. They really wanted me as their new head of Engineering. The second, Avid Technology, offered me less than half the salary of the first company, and no bonuses.

Avid was a fledgling company, not strong in management but growing fast. I knew that I could add value to their haphazard way of doing engineering. They needed considerable effort to keep them profitable and I saw myself doing a lot of good there. I admired the man who would be my immediate boss and liked the way he thought.

There were other advantages as well. Avid was doing video editing—their software and hardware was fun! Although the position was VP of Software Engineering, I'd be responsible for all the drivers and the circuit boards that went into standard hardware—usually the task of a hardware executive. Drivers are written in the kind of low-level software language that I liked to work with. While it was harder to write in these languages, they afforded the coder more ability to directly manipulate things. And Avid was also closer to home; with my youngest now in college, I could spend more time at work and still enjoy a shorter, easier commute.

I took the lower paying job, much to my husband's chagrin, although he didn't push me either way. I thought I'd stay with this company for the balance of my career. After my long stint at Digital, I was ready to settle in again and happily reported to the Senior VP of Engineering, Ozzie.

When I arrived at Avid to take my new position as VP of Software Engineering, the company was just being divided into groups organized to maximize profits. Three business units had been established: a Traditional Unit, which sold the software we already had, based on Apple hardware; a PC Unit that moved the technology onto PCs; and a Broadcast Unit that moved the technology to UNIX servers with the Traditional Unit's product as the clients. A core-engineering group provided 90% of the software to all these business units, each of which had its own small group of engineers testing for their customer base and adding value to the software provided by the core group. The company did no testing across groups.

My job was to manage the core-engineering group; it was a cool job. Ozzie knew what he wanted and laid out four objectives for me to accomplish in the following 18 months:

1. Grow a core group of up to 200 of people who are healthy under my supervision and do productive engineering. Create the "core software" for the company.
2. Establish a Quality group for the company. Put it in place so that it impacts the results of the company.
3. Be responsible for the quality of people who are hired.
4. Grow Avid Technology to have an environment of successful cross-platform development.

I envisioned that if I could pull this off, Avid Technology would be able to do wonders.

I was given about 17 employees and told to hire to accomplish my four outlined goals and to focus on making the core engineering uniform across all the business units. I expected that I could bring this core group up to about 200. Business units would then only need to hire a few engineers to test the unique parts of their product and to specifically enhance products if needed for their profit. The whole idea was elegant and I loved buying into it. I didn't realize that the concept was only in the mind of my boss, and not necessarily in the minds of the rest of Senior Management or the Board of Directors.

Avid was fun. The people that I was given were writing the kind of software that I loved: drivers and low-level software—what we called the "media engine"—that allowed Avid to do its video editing. The company was selling 13 distinct "products" to sports teams, high-end video editors, and creators of commercials and TV shows. From day one my goal was to have that media engine and all that went with it running not just on Apple hardware, but on UNIX systems and PCs, and be used interchangeably across the multiple products. We even had an in-house camera product with terrific technology that would use the same media engine, although it ran on a UNIX base. The software engineers working on the camera product reported to me.

LIVE TO BUILD SOFTWARE, AS USUAL

This is how we grew the business. Everyone wrote software frantically all day, and the working software was folded into what was our core software at the end of each day. I had a team that built tests and tested this to death. By the following morning, they could tell what of the new software "broke" our existing good working software, and what had to be fixed before it was accepted into the core pool. This kind of process hadn't been done at Avid before. The result was that soon the camera team on UNIX and the PC team and the team running on Apples were all running one clean core of software. Quite an efficient operation!

The Traditional Business group sometimes complained that they couldn't get clean copies fast enough running with the capabilities they had asked for. They started adding lots more to the "top" of our product to satisfy their users, which required more testers to ensure their product was working correctly, and actually caused more diversions from the core and more expense to the company. This ultimately diminished the purpose of a core team, but this Traditional Business Unit had been accustomed to being the only one in the company and having everything always done directly for them. The PC and Broadcast business units just accepted our product and pretty much used it, making no changes and few additions. These two new business units loved the idea of a core group; while we were completing a lot of work they asked us for, they could move faster and focus on their customer base.

KNOW HARDWARE IF YOU WRITE LOW-LEVEL SOFTWARE

During the reorganization into business units, two groups had been established to service the entire company across business units: a new hardware group and our core software group.

Avid Technology had been selling its video editing systems by building the software in-house and contracting out for boards made by TrueVision, a company from Indiana that specialized in such development. The boards went into Apple Macintosh computers to collect streams of video, and the customer could then edit that video on a Mac. Apple computers were a wonderful platform for our users—they were so good with graphics, were easy to use, and were much more accessible in scale. Moore's Law reminded us that our processing power had been doubling every two years—editing video could now be done on a small machine!

To that computer/board combination we added our software to do video streaming, plus a whole lot of software that talked to the board to bring in the video, edit it and use it appropriately. My group wrote those drivers to talk to the board, and as our two new business units began utilizing these same boards that were put into PCs and into the "clients" in the broadcast product, we modified our drivers to take advantage of each of the computers we wanted to run on.

I thought of the arrangement with TrueVision as a "just in time" board development group. I got to specify what we wanted them to build and worked with them daily to optimize it to the specifications our software required. That way we ensured that each of our business units could sell the hardware/software product with guaranteed confidence in the software that was driving all the hardware. And because my group did this hardware "sourcing" from elsewhere, Avid's new internal hardware group could focus on new products to grow our business, specifically in the short term, a new camera product. We interacted with that group—an energetic bunch—in other ways as well: my software people connected with their hardware people for one particular project, conducting joint reviews and joint design sessions. I loved all of this; I had always done well making software work with existing hardware.

In many ways managing the core group was like managing a software team that performed like a hardware group. I was determined that my core group would know the hardware for the working system. That knowledge was a natural result of us understanding our business, talking daily to our TrueVision board providers, and clearly defining what we wanted on the video boards. Just because we outsourced the boards to another company, didn't mean we would neglect knowing our hardware as intimately as though it had been provided in-house.

TrueVision employees appreciated and respected us for knowing their boards as well as their engineers did, and the company became delightful partners. (In fact, some years later on the day I was set to leave Avid Technology I got a call from the president of TrueVision, who asked me to come straight to Indiana to interview for a job with them.)

VP OF WHATEVER THE COMPANY NEEDS

And what of my job as VP of Software Engineering? Beth, our HR person, once described it to me so well. Someone had asked her what a VP does in our company and she got a picture of me in her head. She said, "Well, Bev, I saw you this morning in the conference room talking with a group of customers and you owned the room. I was so impressed. I saw you after 5 this afternoon, and it seemed you were sitting on the floor with a couple of engineers counting screws. Now I don't know what else you do, but your job sure has variety!"

My job did have variety. I did whatever had to be done to get our products out the door. I needed to know what all the engineers were doing so that we could optimize their work and maintain leadership in this growing industry.

The Broadcast business unit, new and not yet bringing in revenue, was intent upon selling a gigantic server with up to a dozen "client machines" attached. When a stream of video came in (live newscast info, for example), any or all of the 12 client machines could modify, cut and patch the stream of video in real time, and then send it out to be broadcast in very close to real time. We were doing this by taking in digital, not analog, signals and this was new to this industry. News stations loved this idea.

When I joined Avid Technology, the broadcast product had already been sold, but not delivered. At that point the company had a large machine (made by Silicon Graphics) to be used as a server, and only one client could be attached without the system dying. Video streaming and editing absorbed machine cycles, and I believed that the engineers were going about the process without foresight. But redesigning how Avid received streamed data on 12 clients attached to this huge server wasn't an option. It was too close to shipping time, and we had to make the product that was sold "work." So we *did* need to get those 12 client machines streaming video all at once and working against our server.

Video consisted of a massive amount of data going over a small wire, and we had physical limitations even with an expensive server and the latest technology.

My background in client/server technology indicated that we shouldn't be sending all this video out to the clients and then back again to the servers. We could have just kept the high-resolution video on the server, sent something with low resolution to the clients who had to edit it, and then send the changes back. We could then stream the final video to the TV station from the server. But that wasn't what was sold, and so my head was full of ideas for making *future* versions so much more efficient. It was so easy to be thinking ahead.

The core team was responsible for modifications to the core software that the Broadcast group needed. We had to do some redefining, determining, for example, how many of those 12 clients hanging off of the server could actually modify the same content at the same time. We finally got it all working and had happy customers. The model, though it worked well, wasn't very profitable during the time it was being configured.

I'd learned from my former boss, Greg, that a good software manager has to develop a sense of knowing what people *want* to buy. His view was that a company couldn't grow using the "everything works with everything" model; that approach wastes engineering energy. At Avid we needed more emphasis on developing our unique competitive advantage, so we steered our efforts to give that to broadcasters via our new digital systems.

Marketing was selling expensive systems without a lot of profit margin, and little in the way of strategy. For example, Marketing hadn't determined how to sell digital systems to customers whose unions were planning to push their companies *not* to buy our new technology if their members knew how to work only on analog systems. Getting our foot in the door in each news station across the country was an unanticipated challenge. Thus, the Business Unit had a slower take-off rate than Avid's business models had predicted. The company obviously wanted the Business Unit bringing in profit, but getting that group's product working and salable was a big step. Again, testing was the costliest and most difficult part of this endeavor.

Making the Business Unit profitable normally wouldn't have been the function of core engineering. But efficiency was the byword of the core group, whose goal was to keep as much of the product common, test it once, and then send it out to be sold in its 13 Avid Technology products. The Broadcast Unit had big servers with all of these attached clients, and broadcast products just stayed in our core group, so the Business Unit didn't have a lot to add. The advantage? All our core software was running on the Broadcast Unit's UNIX server, on the PC clients or on the Apple clients. That meant the Broadcast Unit could use the core software, add little else and yet have a solid product that was ready to be sold. The game plan became: only add value if customers make it a good business proposition. Save the company big money by doing it efficiently once.

Chapter 10

*It's not what you gather, but what you
scatter that tells what kind of life you have
lived.*

—Helen Walton, wife of
Wal-Mart Founder, Sam
Walton

THE SOUL OF THE COMPANY SOFTWARE

The business model worked. My core group at Avid Technology
became a lean mean machine. Company engineers enhanced their
software daily. Our group provided clean updates to all the groups
working on top of the core product. By now daily updates included
much more than the media engine—but when we provided a working
version to other parts of the company, the same core worked on the
UNIX machines as on the PCs and the Apple systems.

We also met with the PC business unit, and they were just so fun and
easy to work with. They pretty much took our product intact and sold
it—they didn't need enhancements to make a unique video-editing
product on the PC, they just needed the Apple product running on a PC.
Over time they, too, tweaked it to make it best for their customers.

Daily I worried. How could we maintain our killer media engine and
such great video-editing software that continued to give us an edge on
the competition that was growing around us? We were the first
company to do video editing, but I worried about the Apples and the

Microsofts of the world folding free video editing into their operating systems, and cheaper systems that were "almost" as good as ours being provided by others on PCs. We were being paid a premium because we were the best, and that meant that we had to continually be better. We needed to grow the capabilities of our systems, but not spend money adding bells and whistles that had no intrinsic value. We had to be *such* a cool editing system that we could still charge a premium.

At Digital, I had watched our spectacular mail system no longer command a premium once PCs folded mail into their operating systems and offered it for "free" with their systems. I knew it wouldn't be too long before that could happen with video editing, and we had to be prepared.

I kept hiring good people. Sometimes I got lucky, and great people who'd worked with me before called and wanted to know if I had interesting work for them. I got one of my best engineers that way. I couldn't even match the salary of this engineer, who I knew was a fantastic engineer. But I could give her interesting work. She decided to take a cut in salary and come work for me. I put her on a hard project called "multi-cam." What if four cameras were *all* shooting the same movie from different angles? Could we see all four views simultaneously and, keeping it all in sync, "cut and paste" between them, pulling out only the action we wanted from each of the four views to make one outstanding video? After *much* work and hair pulling, we proved that indeed we *could*. Customers loved the feature, and our engineers received an Academy Award for Technical Excellence for that multi-cam feature. Because we loved doing the best, engineers were challenged and stayed.

And for me, I'd pulled an engineer into the company who was very talented, enticing her with the chance to do something difficult that no one else had done before, and she did it. Entice the best engineers by providing the best chances for doing challenging things—I loved it. I wished I could have given monetary compensation but, like me, many of these engineers were okay with "enough." They didn't need to be the most-well-paid employees as long as they were given interesting work.

While experiencing all this satisfying work, I also felt a special sense of gratification within my family by seeing my children thriving during this time. Ken and I had guided them through high school; each was then able to flourish at his or her chosen university and graduate. I found that my life had some limited social aspects through church and through my husband's interests, but my primary points of focus in life were still my husband and children and my work. That's what filled it up: family and work. Finding additional time or energy for friends or other interests wasn't in the cards yet.

BUILD IT OR BUY IT

We engineers were working hard building and maintaining our products, but it was 1992, and Avid Technology had begun looking at products of other companies that might complement our offerings. Avid bought a couple of these new little companies, and one of them was assigned to me. What a gift! This was a small Midwestern company that created video "clip art," and that was owned and managed by a software genius. Orv cleverly came up with ideas and then made them happen with software. When I visited his company and studied their software it was flawless and beautiful; I was so impressed by their clean code.

Avid Technology's software was a hodgepodge of spaghetti code that had been built quickly and enhanced continually, and when I inherited the code it was not worth rewriting. It was a disaster to enhance because it wasn't easy to decipher. But this little company in the Midwest had an owner who didn't *let* people add code to his pristine code unless it followed his rules and was as clean as what he wrote. Orv's company consisted of about 20 engineers who all followed his guidelines, and his rules for writing code made sense. Ollie still controlled that code, and I loved what he did. To our company that meant Orv's code "did what it was supposed to do, was easy to maintain, easy to enhance, and cost less to test"—all the things that make code good business.

Avid Technology should have adored this genius, and while we gave Orv ample reason to stay with us for a while after we bought his company, the *real value* of that acquisition was Orv's *mind*—and we needed to tap his genius to feed other parts of Avid. I didn't stay with the company long enough to do this, and once I left, they ignored Orv's significant value. Because of that, Orv moved on to do other things. It's sad to lose some of the company treasure without the people in charge ever recognizing the glittering jewels for what they were.

THE JOB ISN'T JUST ENGINEERING

Software is interesting to develop, but it is a unique business. Software is invisible. It doesn't obey any physical laws. There is no manufacturing production—we sell "prototypes." Software absorbs functionality, and there is no limit to expectations customers and management have for it. I loved the challenges that writing and working with software offered me and my engineers.

When I arrived at Avid Technology the company had few (even junior) managers. In the beginning, my engineering team looked at what we could do that day, each day. I constantly studied what was *not* being done in our plans, evaluated alternatives and costs, and recommended solutions. Most of what our core engineering group did could be used across business units and the new business units loved what we were adding and could sell it. Some of our core group's work was a result of directives from the individual business units, which could then be offered to all business units through our common software.

During my tenure at Avid, I was able to grow some great managers by giving them hard jobs, effective tools and the incentive to build great software. Hockey star Wayne Gretzky best stated what I tried to instill into Avid Technology management: "I skate to where the puck will be." With this method of efficient engineering that I championed, Avid Technology grew from an $11 million company when I arrived to a $52 million company the first year, then $200 million the second year, and a $400 million Digital-Equipment-Corporation-sized company at the end of the third year.

That was enough for me. Even if I couldn't instill Gretzky's concept in the trenches. Because the *first* thing each business unit wanted my group to build was what they'd already sold – not something new.

The Avid Technology Board of Directors was struggling with the company's original CEO, believing that a CEO who started a business wasn't the best person to maintain and grow the business once it got to a larger size. They were frustrated because the two new business units hadn't grown as fast as had been expected; they felt that the Traditional business unit shouldn't still be carrying so much of the profitability of the company. With their high expectations not being met, an insidious "throw the rascals out" mentality began to take hold among board members.

I had been working hard with engineering, and not watching the meetings at the top of the company. I respected Ozzie and thought, as my boss, he'd cover my group well in top level meetings, because he was rational and smart. Even when Ozzie **told** me periodically that strange things were happening above us, implying that things weren't going well and that he wasn't sure how it would turn out, I pretty much ignored those red flags and kept working. My fatal flaw in this company was not being "political" enough. At least one of the business managers didn't even understand what our core software group did, and I depended on the Engineering VP in that manager's business unit to get those details across to him. I shouldn't have.

The board showed their displeasure for failed (even though unrealistic) expectations by instituting a cleansing of all of the top management and then putting in new blood. I was caught in this cleansing, although I wasn't immediately moved out when the CEO and others left.

My best advice came from one of the executives, who said, "Take a package and get yourself out of here. This company will *not* be the exciting company that you came to manage. Move on." Well, I had planned on settling here for my career, but things had quickly become chaotic and the new management didn't give me any desire to stay. The new senior head of Engineering said, "I don't know hardware, so I can keep the hardware manager, but I know software—so why do I need you as another software manager?"

That was such a ludicrous question that I chose not to answer rather than have to explain it! I understood I was cutting off my nose to spite my face. But this person didn't even realize that I was responsible, from outside sourcing, for *all* the hardware being shipped as part of every current product that the new senior head of engineering had sold through his business unit. The real business of the company was building software systems, and that had been my specialty.

I thought back to what I had been tasked to accomplish when I joined Avid:

1. *Grow a core group of up to 200 of people who are healthy under my supervision and do productive engineering. Create the "core software" for the company.*

I had created core software across products for the company. I was proud of the software development team I had grown—186 people who reported to me. Efficient and effective, we created software systems that did video editing better than any other in the nation at that time.

2. *Establish a Quality group for the company. Put it in place so that it impacts the results of the company.*

Our testing of our products was very automated, led by strong engineers who were determined that no product would be shipped without being polished and ready to be used by customers.

3. Be responsible for the quality of people who are hired.

I was *so* proud of the engineers and managers we had hired hired—I'd gone from 19 engineers and managers in my group to 186.

4. Grow Avid Technology to have an environment of successful cross-platform development.

Our products ran seamlessly on Apple systems, on the UNIX operating system, and on PCs. They ran in our little cameras and they ran on huge SGI servers. We excelled at cross-platform development.

I was proud of all I'd accomplished, but I had to think ahead now. Could I and my engineers thrive under new management? The answer was straightforward: not and continue with the fun that we'd been having. The politics of the company was about to decimate engineering—that common base I'd worked so hard to build. It seemed the right time for me to move on. I did my best to help the company define those people they would need as they reorganized, because management didn't know the value of those engineers.

For myself, I was learning that I did well when I was building products that had never existed before—products that called for unique design and development—even though I didn't know exactly what the end product would be. I needed to find a place where I could continue inventing exciting "stuff."

As I progressed in my career, my experience confirmed that the single greatest thing a company can do to retain employees at whatever level is to make sure those employees have reason to respect and value their immediate management. I made a practice of evaluating each person in my groups, and if two people weren't meshing in their manager/engineer relationship, I would reassign them within my company. Supporting the "I value my boss" mentality made such a big difference in attaining and keeping engineers satisfied with their jobs. Respect your boss or find another job. I was beginning to understand that I needed to follow that advice, too.

I knew my software at Avid Technology would last about nine months to a year without anyone doing any updates, so I was ready to leave. Eleven of the 13 engineering managers from my group were gone within two months after I left. Ironically, it was the new management that actually lasted only nine months, and top management turned over again. I had learned at Digital Equipment Corporation that when you left a company, anything you had done was considered passé, and they would just move on. This was also true at Avid Technology. But this time it was more painful, as the ideas that could have made Avid a $600 million company within one more year were so strong I could *feel* them. But once again in my experience, the powers-that-be (this time the board of directors) didn't know what the company had or how to take advantage of it. Ten years later they were still making about $400 million per year.

Avid Technology was good experience under my belt. But it was time for me to make my mark somewhere else.

TIP #9: STOKE YOUR FIRE.

You get stale if you stay in the same job for too long. Whether you take a series of jobs with different companies or move within a company, try to make each new challenge stretch you. Find ways to make yourself think outside the box.

Chapter 11

*I slept and dreamt that life was joy. I awoke
and saw that life was service. I acted and
behold, service was joy.*

—Rabindranath Tagore,
philosopher, author,
songwriter, painter,
educator, composer, Nobel
laureate

MOVING RIGHT INTO HARDWARE

My time at Avid Technology proved to me that I could institute strong
methods for efficient and effective software development which would
affect the company's bottom line and at the same time keep engineers
happy because they could concentrate on interesting development. The
question for my next endeavor was: Could I extend my methods for
efficient software development to include hardware development?

It was 1996. Hardware chips were becoming very complex, and
managing the smart engineers who crafted them was an increasing
challenge. Strong engineers typically danced at the edges of their
inventiveness and at the far reaches of their imaginations. This
exponentially increased the difficulty of their efforts. And it made
managing these engineers—understanding them, supporting them,
inspiring them—very tricky.

In my job search, the work of Number Nine Visual Technology caught my eye—literally. They produced 2D graphics chips, and graphics boards (some with their own chips and some with chips from other companies) that fit into a PC to provide all the graphics for that PC. At the time, Number Nine 2D graphics cards were in all Dell computers, and were also being sold to other companies who needed graphics abilities in their systems.

At Number Nine I would have under my management umbrella all the testing of our systems, an area not well developed. I would also manage the customer support group; the company had a lot of systems to maintain and provide support for. I knew I would have to put a lot of my soul into making customer support efficient, otherwise that area was going to eat the company alive.

But the main reason it was going to be exciting to join this company as VP of Engineering was that they were working on a 3D chip, and the heart of that job was managing the hardware, software and board engineering groups to deliver that new 3D chip on a graphics card that could be used by anyone in the industry. I wanted to keep working at the leading edge. Number Nine Visual Technology needed to create this new technology, get it to market first, and make sure it was good enough—all while enhancing its maintenance processes and continuing to make 2D products to keep the company in business until the 3D product was ready.

My added value as engineering manager focused on three points of efficiency for the software team: writing drivers for the hardware, developing in software anything that didn't work quite right in the hardware, and creating software that would make use of the fantastic graphics capabilities that the chip offered.

My engineers were very good. We had hardware teams developing the new chips to move us forward, software teams enhancing and supporting current products, and a board team developing efficient boards using our chips or the chips of other companies to be sold with our name on it as a "graphics card." In addition, the software teams had to anticipate what software would be needed for the 3D chip being designed. I managed these groups carefully, making decisions that freed the engineers to spend their time on the inventive work we needed. I streamlined the way we tested our product, and protected the engineers from being pulled by other parts of the company to do fixes or enhancements that might be interesting, but that wouldn't get the essential core work done.

A 2D COMPANY THAT NEEDED TO BE A 3D COMPANY

At this time 2D chips were very competitive—we had to be fast and efficient to play with the bigger companies. More than that, we had to look ahead: 2D technology wasn't where graphics card companies needed to focus. At that point in time, just game companies needed 3D technology, but increasingly, more and more computer companies felt they needed to buy 3D boards. Number Nine Visual Technology needed to have a presence to remain in the game.

I remembered times in the past when the industry had been dramatically and drastically changing, when people in charge had to make critical decisions for the future of their businesses. I thought of Digital Equipment Corporation. In its heyday, Digital had thrived because they built mini-computers when most computers were huge. But the company had not been agile enough to adjust quickly when the technology moved fast and those small, powerful computers got cheap. I also thought of my professor at the University of Dayton, who had been a genius for seeing a new field emerge, but then couldn't recognize when that field was moving so fast into small technology.

The hardware team that was developing the new 3D technology for Number Nine was the hardest to support with management. They were a tiny group, and our competitors were throwing a lot more money at this problem!

At first, a critical member of Number Nine's top management, Bob, visited the head of the 3D hardware team often, so I backed off, cleaned up the rest of engineering, stayed clear of hardware and left them to do their own thing, thinking that they were about finished and I should keep my hands off. But they weren't "about finished." I assumed that, through his many visits, Bob wanted to keep a close eye on their work. As each day went by and no "end date" was in sight, however, I realized I needed to work with the team.

Hardware had agreed to do something very complex and difficult. Smart and savvy engineers though they were, they also were so stretched that it seemed they needed every minute of every day to develop. I didn't want to take their time by talking to them. The company so badly needed what they were building that I hesitated to bother them to review work or take any minutes from their day. But with the leader of the 3D hardware engineering team holding so much in his head, and the company needing to rapidly move forward on this project, I knew I had to figure out how to help to successfully finish the design and development of the chip.

Then, in a meeting with senior management from Dell computers, one of the executives made a remark that made a light bulb go off for me. He said he appreciated Number Nine's cleverness in heading their engineering group with someone well versed in software development, because that process was so applicable to manufacturing chips!

I hadn't realized that knowing how to develop software was such a powerful tool in my back pocket for the hardware teams as well as our software engineers. During my career I'd never just maintained software or developed simple software applications. I had always lived in the middle of hardware people, building operating systems or compilers or plant management systems or video editing systems that needed software and hardware to seamlessly work together to produce products. This Dell manager had inadvertently helped me focus my own talents to help the hardware teams. I suddenly felt we had a chance to put Number Nine's name in lights in the graphics world.

I began working with the software teams and the board team to anticipate what they were going to be getting when this 3D chip came out. We knew the designs and what might not quite work, which software could then build to cover, and we could predict what was needed for board work. Trying to speed up these processes would shorten the time to market for this 3D chip.

Working with the hardware teams, I purchased an emulator for Number Nine. I worked with young, inexperienced engineers to duplicate in the emulator the main 3D chip we were building. We had to do that without using up much of the time of the chip-development team, the only ones who knew how the chip was really supposed to work. I directly managed this little emulator team, since the investment in an emulator was huge, and it had to be of value to us.

Our emulator team instigated brief design meetings with the chip team, because the more the designers could articulate what their chip was doing, the more they would see where the fatal flaws might be, and the tighter the development became. I knew the limited and valuable nature of the time the chip designers had. If engineers got 15 minutes with the design team—that was like "time with God"—they had to make the most of it! I worked with the junior engineers doing emulation, fielding those questions that I could answer without bothering the 3D chip development team. By the time they were ready for their 15 minutes with the chip designers, their remaining questions were focused and difficult, and allowed us to move faster.

Every one of us lived and breathed this 3D chip. My job was to facilitate and to hone all the work of engineering to give the core hardware team the best chance for success. I did my best to hold on to all the surrounding work so the core team could spend all their waking moments on the invention at hand. I even quit renting the building where my customer support team worked and moved them into the main building with my software engineers, thus freeing enough dollars in the engineering budget to open a small hardware office in nearby

New Hampshire. This cut the commute time of the hardware engineers and gave them a better chance of success.

To fabricate the chip we used IBM, which at the time was the most progressive chip builder in the U.S. We chose them because they were good at it, their fabrication plant was in nearby Vermont, and they were responsive. I learned much about how IBM built chips during our many meetings with IBM representatives, getting ready for them to fabricate this chip for us. Fabrication could take four to six months and we needed the chip design to be pretty flawless to get it right the first time through. Our software team had to "fix in software" anything that didn't work perfectly in hardware, and they needed their software to make use of every feature the chip offered. Our board team had to have working boards ready, containing this designed chip as soon as the chip was available, even though the team couldn't depend on it to work just as designed.

During most of the time I worked for Number Nine Visual Technology, I didn't work much with Bob. As part of his upper-level management duties, he often traveled to other countries to find places to mass-produce our chips economically. Traveling was a challenge for him because it precluded him from seeing what we were doing, which was a departure from his previous experience. Bob was used to being more aware of all of development.

That now fell to me. I kept everyone out of Engineering's hair so we could produce strong products. We had a new, experienced COO who was driving business and I just focused on engineering development. We were shipping excellent 2D graphics cards and the company was happy with our engineering prowess.

But when Bob returned from his trips, he was nervous about our 3D chip. Once I found him up in New Hampshire smoking with my engineers, telling them it was okay if this chip didn't work and okay if they slipped the schedule—he'd find money to develop another one. He was trying to make them feel good, but that attitude didn't make any of us feel good. I was confident of this product. The emulator had helped us weed out problems and develop some software to give us a head start. We had the board plans underway. We had a *fantastic* chip and were trying to adhere to the schedule and ship as planned. We were all living for the success of our 3D chip.

WINNING WHILE I WAS LOSING

Like any good member of upper management, Bob needed to look at the bottom line, and apparently had already decided that if this chip didn't work, he would fund another follow-on chip, or at least find funding to fabricate this chip again with needed changes. What else might help his bottom line? He was willing to work with Marketing to price this 3D graphics chip for quick sale to get on with producing the next one.

Bob's lack of confidence in our 3D chip was subtle, persistent, ultimately damaging and a surprise to me. If he didn't expect this product to take off, he hadn't shared any of his concerns with me. I didn't know he was already looking at alternatives if the chip didn't succeed.

Meanwhile, engineering had their heads down, were focused and remained diligent. Once the design was declared done, the chip was fabricated. I drove with Number Nine chip designers to Vermont a number of times to work with the fabricators, and finally, after months, we got the working chip back. As fast as possible we put our chip on a board and got the software working. Finally we were ready to ship our product!

It was only then that I allowed myself some medical leave. I'd scheduled some necessary personal surgery around the chip completion, postponing a hospital stay until after the 3D graphics card could be released for shipping. (Yes, one can be very stupid about personal things when in the throes of birthing a new product!)

From the hospital, I called my engineering team every day the first week to check on the shipping status, and then from home I worked with them by phone each day. During those first two weeks after my surgery I felt a great sense of fulfillment while pondering our engineering accomplishments. In fact, I was euphoric. Engineers and marketing staff were planning to take our new 3D graphics card and show it off at COMDEX—an annual computer-products trade show that we'd worked so hard to be ready for. We'd made it!

Feeling the success of the 3D card was short-lived, however. During the third week after my surgery while recuperating at home, I received a call from Human Resources letting me know that my position, VP of Engineering, had been eliminated. And I learned I was not alone; the COO position had also been eliminated. This was the company's solution to optimize the dollars needed for the next chip that would keep Number Nine in the graphics business. So I no longer needed to call in each day after all; my job was finished.

COMDEX, the darling of 3D technology makers, was happening the following week. In past years, not one of Number Nine's previous graphics cards had ever won a single award at the show. But this time our new graphics card won *seven* awards—all that COMDEX had to offer! Even before the presentations, I'd already felt great satisfaction with the work of our engineers. We all knew Number Nine built a fantastic graphics card; finally, though, we had the awards to prove it.

Although Marketing was ecstatic, unfortunately the awards could do little to advance the product because the company, not expecting such success, had priced it too low. A small thing, seemingly, but with large consequences. Number Nine wasn't set up to selling our 3D-chipped graphics card and touting it as they should have—and thus missed the opportunity to take over the graphics world. Shortly after that, other graphics companies including ATI Technologies and Nvidia were able to produce very competitive 3D chips. They prepared the rollout for their products and had the support in place, and those two companies in particular became the dominant players in the industry. Number Nine

Visual Technology lost its lead before it could capitalize on even *having* the lead.

The business was so competitive that this little company couldn't hold its own while developing yet another chip. The 3D graphics technology of Number Nine's first 3D chip was purchased by the clever engineers who had invented it; they used it for a number of years in a company they formed around the technology. These engineers, and all of Number Nine, could be very proud of this breakthrough technology.

I, too, was very proud of my engineers. Even though they were not "my" engineers anymore.

Chapter 12

Some people want it to happen, some wish it would happen, others make it happen.

—Michael Jordan,
American professional
basketball and baseball
player

NEXT STEP, A BIG STEP

My body was not all that needed a bit of healing following my surgery. I was still recovering somewhat both mentally and emotionally from the stunning news delivered so casually by phone that my position at Number Nine Visual Technology had been eliminated. I really had not seen that coming.

But even though the doctor suggested I spend six weeks recuperating, about three weeks into my sick leave, I received another call—a job opportunity—and began thinking about work again. I'd been lucky to have had several offers when I'd taken the job at Number Nine, so before I was even off their payroll, I was asked to come look at one of the companies I hadn't chosen when I joined Number Nine. This company had something right up my alley, they said, and they wanted to talk to me. (How about that to boost one's spirits?) Well, even though I hadn't sent out my resume yet and no one knew I would shortly be on the market, I was ready to talk! It was 1997.

I spoke with Mitsubishi Electric of America in Cambridge, MA and became intrigued enough with their opportunity that I agreed to get myself to an interview a mere four weeks after my surgery. Just driving to that first meeting was physically so taxing that I knew I shouldn't be driving yet, and moving around was a pain, but the first person I interviewed with was compelling and described a proposal that appealed to me. I had them set up further interviews.

Other offers I'd had when I chose Number Nine were general manager positions, with upwards of 300 or more people to manage. I considered them briefly at this point, but I'd already done that kind of managing.

I kept thinking about Mitsubishi's opportunity. My curiosity piqued, I wondered whether the visions they shared with me could become reality.

They had a highly ambitious idea in their research lab to build an inexpensive, readily available volume-rendering board that could go into a PC. Volume rendering was a set of techniques that allowed a user to manipulate a 3D image on a 2D display. It was considered the "ultimate 3D."

At that time, when the industry built a 3D chip, they did something like wrapping a wire mesh around a 3D object, layering smaller and smaller pieces of this "mesh" closer and closer together, giving the user the illusion of three dimensions. But, like the process of papier-mâché, the 3D graphics cards just covered the outside of the object, although the result was like viewing in three dimensions. There was no way to see what was *inside* the object.

Mitsubishi Electric wanted to keep track of every dot that was inside of that entire 3D object, using "voxels" (volumetric pixels) and not just the pixels that we see on a 2D computer screen. Instead of storing a flat picture or a square from the data taken in, Mitsubishi wanted to store a *cube* and every dot that was inside that cube. That is defined as "rendering" a volume. The end result would be to render for view the object's inside *and* its surface—but more importantly, to see it from any angle, from any direction, with any specificity. But that wasn't all. Mitsubishi then wanted to provide all that information to the customer *in real time*. The company had proven the viability of the concept; they had worked with a university in New York to successfully build a prototype that was able to do this volume-rendering by using something like 16 chips on a board. Quite a challenge.

My mind spun with the possibilities: voxels instead of pixels, viewing the substance of the inside of an object (or a person!), processing in real time—this was all *very* exciting. But I needed to consider: was it *too* ambitious?

This product would never be a competitor to Number Nine's products, or ATI's or Nvidia's or other graphics companies, because this wasn't a 3D graphics board. It would cost too much to be mass produced. This chip would be used commercially in a CT scanner, or in places that needed more capabilities than standard graphics cards could begin to offer. To put this capability into a board that could then be put into a standard PC, that PC product would need to have "high end applications." Medical applications, for example. This was not yet done—not in a financially feasible way for either producer or consumer.

Mitsubishi's question to me was: Could I join their company, create a group and build a product from this research that was cheap and effective, and would create sales worldwide?

I had to think about this opportunity differently than others; it would probably do my career no good. I wouldn't be going up the career ladder in Mitsubishi, moving to Japan to take higher roles in management. They kept their American and Japanese branches separate, and I might never achieve a higher "level" in the management chain at Mitsubishi. The job they were offering would allow me to give only a small number of direct reports and therefore meant I'd have less management responsibility and complexity than I was used to—not a strategy that would typically serve to advance one's management career. There were other, more career-enhancing job options, I was sure; I just hadn't searched them out yet. But then, maybe I felt like I didn't want to.

I'd just suffered two big hits. One, of course, was the major surgery. But losing a job that was so important to me required recovery as well. During my six weeks of recuperation, I'd had time to think about what was important, and how much control we really have over life and the work we do. Fighting illness and psychic injury remind you that life is short. Now, on the other side of those unexpected battles, besides contemplating Mitsubishi's question to me, I had to answer my own question to myself: What did I *want* to do with my life that was significant and fun and challenging?

From that perspective, what Mitsubishi Electric of America offered me was a chance to apply my smarts and experience to hardware issues and building chips, and to apply my software background and management ability to create a product that could do this particular kind of "volume rendering"—something that had never been done before!

Could I "productize" their idea? The challenge was exhilarating to contemplate. I did have experience in building board s and chips and software. Mitsubishi had been correct about this job being right up my alley. It fit my skill set and definitely piqued my interest. But, practically, before making a decision, I needed to study the people and resources I was going to be given to make this happen.

The existing team from research that I was to take over consisted of six engineers, who I interviewed and determined were very good. In addition, I'd have to quickly find and hire about 20 more engineers. In total, we'd be allotted about 18 months and a six million dollar budget. Since it normally took four to five months to fabricate a chip, that meant a little over a year to find engineers, then design, create and fabricate a chip and build the software to make it do what was desired. I knew that Mitsubishi Electric of America usually took about four years to build a chip, so even the company had to be skeptical of this little group's ambition. I'd been used to trying very difficult things while managing hundreds of people. That situation might engender in someone more faith in a positive outcome: the more good engineers, potentially the better the product. But in this situation, there would be just six plus those I could attract quickly.

So that became another key question for me: How did this little team come by its great confidence to accomplish such an ambitious task?

I found the answer in the man who made me the offer. Kevin had tried to hire me before and was well known in the industry; I trusted and respected him. He was about to retire. He wanted this project to succeed, and wanted it set up before he left. Though Kevin obviously would not be my boss, he told me who his successor was: a big, big name in the graphics field—and Kevin even let me talk to that gentleman by phone. As an international company based in Japan, Mitsubishi Electric of America valued people who had made names for themselves, and this man who would be my boss fit the bill. I could see it was a high quality operation.

Once again my decision came down to the value I placed on the work and my co-workers. I loved the interesting technological challenges this job presented. The people I was going to work with were smart, they were good people, and they were motivated to do this! It would be like working for a start-up, but with a big company's backing.

And so I took the challenge. I became VP of Engineering for the Volume Graphics team at Mitsubishi Electric of America.

FORMING A TEAM

We spun our team out of the existing research group and rented the floor below them in Cambridge, MA. Having limited money, we didn't form a board group, but outsourced that work instead—I'd learned that from working at Avid. We hired one man part-time who would do our board design and we found a company who would build boards to spec as we received orders from customers. No inventory in the closets for us! That I'd learned from Number Nine.

I needed a software team and a hardware chip team. Knowing I'd need to budget for marketing and sales as well as engineering from the money I'd been allotted for the entire project, I determined that I couldn't have more than about 20 engineers, total. So we formed our group from the bottom up, keeping in mind the talent needed to accomplish our goals.

Gordon was one of the six engineers from the research group. He was a smart and enthusiastic generalist, the inventor that could oversee all of the work. Ian was a newly hired, fantastic engineer whose forte was taking something very complex and designing it with simplicity. He had recently been hired for his ability to work smart and to build high-quality chips, so he was the natural leader to design the hardware chip. Ian had taken this job for the same reasons I had taken mine—not for career enhancement, but because of the challenge of this technology! Both men, chip designers who had an eye for the possibilities, were driven to do this work. Art was the third of the existing engineers, and he had managed hardware teams, so his expertise was different but crucial.

Our six "originals" from the research group included one more hardware engineer and two very good software engineers. I was lucky in that one of these software engineers was meticulous about designing clean code and could manage this little software group. Great start. But that wasn't the end of it. We occasionally borrowed one of the top engineers who stayed behind in the Research Group: Gil kindly made himself available for consultations and reviews. That original research team gave our new group tremendous support as we started to define the problem.

No one had built chips this complex. We decided that we would do our best to produce a single chip to allow us to manufacture inexpensively—*one* chip, not sixteen!—and we set that as the goal.

While our six initial members began defining the scope of the project, I began to ramp up the team and get other people on board. I was allowed to go anywhere in the world to get the best people, but I couldn't get very many, so they had to be very smart, very competent, and capable of working well together. I had to interview fast and find just the right people or we couldn't succeed. Our time frame was too short. We advertised, searched, used recruiters and did whatever we could to grow our team—even for a while using a contractor Mitsubishi had on board.

And what wonderful people we were finding! One developer from a firm near the Mayo Clinic in Minnesota had built a volume-rendering application all in software (exactly what we were going to build in hardware), which had been shown to save lives. The problem, of course, was that the application had been either *very* slow—often taking more than two hours—or incomplete. Doctors didn't have the luxury of waiting so long for data, even though patients credited their doctor's use of this device for saving their lives. This particular software developer really wanted to be part of a group who could "build it right," and make the application fast enough to be effectively used. He had already determined that his work had to be embedded into hardware to make it fast enough.

We hired a Ph.D. from Switzerland who was a software guru and whose specialty had been graphics and facial recognition software. She was appalled when she joined our team and found that she was the only female on the project so far. She stormed into my office and declared that I should have told her before she left Switzerland that she'd be the only female! Wow! First, I asked her what she thought I was. She didn't count me. So I had to convince her that we weren't finished hiring and that more women would be added to teams. At that time I could find few female hardware engineers, but was increasingly able to find female software engineers among the candidates that we wanted to hire. Personally I was ecstatic that now young women in industry were expecting other women to be on their teams.

BUILDING AND MARKETING

Simultaneously, as the team grew, I was excited to go back to participating in design meetings, calling reviews and being totally involved in every part of developing what we were building. And we were building something big.

Technology at that time could provide a doctor with a requested, specific scan, which, after about two or three hours of calculation would show a single image from one direction. The process was rife with breaking points: the scan could be from the wrong angle and need to be re-taken, requiring another two or three hours to calculate and then produce one more image; the patient could have left the office by the time the scan was ready to review; the doctor could have seen many patients in the hours between requesting and receiving the scan; the doctor could have forgotten the exact thing s/he was looking for. Medicine's use of volume rendering software on the current, small scale was an exhausting, frustrating, ineffective process.

We were taking that process a step further, and we knew that the medical field, in particular, had abundant need for the newer applications we were making.

What Mitsubishi Electric wanted first was to be able to scan and capture images that would allow a doctor to see those images *in real time*, to examine the side, the back, the top of a tumor or a brain or a beating heart as s/he chose, or to visualize areas of the body that couldn't be seen without surgery. To achieve interactive viewing of an object in real time, we had to do it via hardware. Second, we wanted to do it on a cheap PC: add our board to a cheap PC and every doctor's desk could have one. Powerful ideas of high speed and high sales, to be

142

sure, but they met the need only if we built the chip that could handle the "real-time" task and then put it on a working board to add to a PC, which would make it cheap enough for large-scale consumption. Combined, those two factors were our only road to viability with this product. In terms of marketability, however, we saw further than that. We knew we wouldn't have to limit our technology to the needs of medicine. We could sell to airports to scan luggage, we could sell to anyone who needed to see the "inside" of a three-dimensional object.

To help us understand the expanding commercial opportunities, the company brought in their California marketing group to supply the early marketing and sales needs of the group. Mitsubishi management also began looking for a general manager for the project. I'd decided I didn't want that responsibility, but agreed to handle it until they found someone, as we had to move quickly.

I was now working at the opposite end of my software/hardware spectrum. The product that we were going to build had been prototyped by the New York university graduate students we had funded, and we knew of one company that sold a similar all-software product. And just like the software I'd created over the span of my lifetime, the prototype could be built again. But it could *not* go as fast as we needed it to go if it was merely simulated in the software and not designed into the hardware.

Ivan, the CEO of Mitsubishi Electric of America, and Rog, the marketing guru from California, stayed with our team and helped in initial discussions and reviews. The company's executives headquartered in America worked with their counterparts in Japan to form a group that monitored us and reviewed everything we were doing every three months. They were setting us up for success. From the start we had such wonderful direction and help, allowing us to focus our energies designing and moving as fast as we could. We had 18 months to create and deliver a product. Subtracting the required minimal five-month "back end time" that would be needed once our product was designed for fabrication left us 13. No time to lose.

LIFE HAPPENS

The cost of this initial chip would be no more than $6 million over two years, so that had been my working budget since coming on board with this group in late November. By January of 1998 we'd had many design meetings, done much planning and grown the team. Hiring was ongoing. Every meeting was productive, though often we had a clearer picture of what we *wanted* to do than exactly how to get there. With all that on course, I had the added task that

is the universal juggling act of management: do everything possible as well as it can be done—and do it all *within the budget*. I was trying to figure it all out when my boss called me in. Japan's poor economy had caused suffering throughout the 1990s and Mitsubishi was not unaffected by those circumstances. But our company did not lay people off. Instead, the solution was to curtail some projects. Times were tough; I was told my budget might be cut to $1 million.

Well, if that happened, I couldn't pay the salaries of my team and produce at least two fabrications of our chip at about $400,000 each. In fact, we couldn't begin to complete this project with only $1 million over two years, so we might as well just fold up our team's tents and close up shop if that was our new budget. But "if" was the key word. Talks were in process; the final decision would be made in Japan in mid-February.

That meant I needed to go to Japan, stand up for my project, justify my dollars, and see if we could get the full funding put back into the budget. My boss, our marketing manager and our Japanese CFO would go with me, but no one was able to get a clear feeling of what the outcome would be. With the big review happening within four weeks, I began preparing for my presentations—researching the decision-makers who would be my audience, outlining the expanding marketing opportunities for our product in medical, travel and other industries, and understanding the constraints Mitsubishi's executives were feeling as dictated by their corporate and national culture. I marshaled my arguments and my energies.

Meanwhile, another battle unknown to me had been silently underway.

I'd had a routine mammogram in January, and was called back in early February for a biopsy. I was given the news on a Friday morning: breast cancer. My situation required immediate surgery. But I was scheduled to leave for Japan in a week and my project would live or die on the results of that meeting. I could see the irony. But I had to go to Japan.

My doctor gave me two choices. I could have the operation the following Monday morning and go to Japan probably with some inconvenient tubes and a bag hanging off of me, but at least with the surgery complete. Or I could postpone the operation, go to Japan a week from Monday as planned, but *not* stay there the full week. Instead I would come home after a day or so and have the surgery before that week was out. Engineers live their work; I took the second option.

I prepared the next week for both the surgery and the trip to Japan. In the process, I re-lived the events of that Friday several times.

The surgeon had left me a message to stop by her office at 7:30 a.m. before her workday. I could just fit that appointment in; I was already feeling the pressure of probably getting to work late because of it. But hearing the diagnosis, as anyone who has received this kind of news can attest, is a surreal moment. Simultaneously, my thoughts rushed by and time stood still. Then I had to stand up and put one foot in front of the other. Once the doctor had laid out my two options and I'd made my decision for the date of the surgery, I got in my car and headed for work. As though it was a normal day. *Not* going to work hadn't entered my mind.

But I was reeling from hearing the "C" word. Thoughts swirled in my head as I barreled down the highway toward Cambridge. I came out of my reverie when I heard the sound of a siren behind me. I pulled over. "Lady," the cop said, "Do you know how fast you were going?" I blurted out that I'd just been told I have breast cancer and that I was sure I was going whatever speed the car in front of me had been going on this highway and I apologized that I was sure my mind was preoccupied. Then he said, "Lady, if you don't want to die *before* the breast cancer gets you, you'd better quit going over 80 on the interstate!" This policeman was a gem. He brought me back to reality.

When I finally made it to the office, I was immediately immersed in the intense work we were doing, even as another set of thoughts began to grow, hovering in the back of my mind. How do I get through this and still keep these engineers on track? How can I add this to my already busy schedule? I called Ken with the news, but told no one else. I just kept working.

My surgeon made it clear that she thought I'd made the wrong decision. It was not that I couldn't wait a week for the surgery, but that I had my priorities wrong—she felt that I needed to be putting all my energies into me. Into getting myself healed and well. She wanted a full focus on healing, and here I was putting off getting into a cancer support group because I just didn't have the minutes left in my week to attend one. But that was how it was. Support would have to wait until later.

I had to tell my boss and a few key others because I was planning to go to Japan with this entourage, but suddenly had to ask them to make adjustments to our schedule. I wanted my presentations at the beginning of the week, which meant my boss would stay on for the closing discussions and to hear the final conclusions. I needed to head back as soon as I could after the first day or two, after making my passionate plea for my project. The re-arrangements went smoothly.

In the days before we left for Japan, though, I found my colleagues were sort of "uncomfortable" with me—like I was a ticking time bomb and they didn't want to get too near. They didn't know how to deal with me. Well, I didn't know how to deal, either. I had to follow the

146

direction of my surgeon and get myself back on course. Did I panic? Sure. I was panicked all the time: what if the surgery wasn't successful? But my work probably helped me to minimize the "what-ifs," because there just wasn't time to allow slack.

My family was a bulwark of support; my husband and my children were just there for me. Friends from my church who'd survived breast cancer let me lean on them. Putting things in the hands of God from day to day was the way I could go forward.

While packing my luggage for the trip to Japan, I suddenly realized that I had gathered all black and white items to take: white top, black slacks, a black suit jacket— just looking at it, lots of black lay folded there in my suitcase. At the last minute I threw in a red silk blouse, simply because the contents of that suitcase looked too dreary to me. Life couldn't be that heavy.

Our cadre of four left for Japan that weekend, and Monday morning I met with the Japanese group who would be deciding our project's fate. The meetings were full of people I didn't know. I didn't understand who were the influencers, and so I was working "blind."

Tuesday I put on that red blouse and gave the presentation of my life. I asked for the entire six million dollars.

The men in my audience listened to me, smiled and nodded their heads as I spoke. Having worked with Japanese companies in the past, I knew this did not signal agreement, but only that they had heard. So I became even more specific. I was meticulous in describing details of the chip we would build, code-named "RoadRunner," but I was also able to describe what the next three follow-on chips would do and why they made sense. So my audience heard a bit about Condor and Falcon and Eagle. With Eagle, we had planned a chip good enough to take over the industry if all worked well.

With the "big picture" painted, I then went back to elaborate on every bit of RoadRunner as I envisioned it. I offered schedules and specific monthly milestones and all the detail that they look for in projects. I was talking to engineers who had no idea whether or not this idea could even be brought to completion, but I had to convince them that if it *was* to be done, we were the ones with the expertise, drive and commitment to do it. By merely attempting the RoadRunner project, I already knew our group was tackling something very difficult; I had to be fully knowledgeable about *how* we could complete our task, and I had to make that very clear. Whether or not these men would allow us to proceed would likely be based on my attention to detail. I made sure my charts and explanations showed exactly what we were going to do with their money.

My Japanese counterparts and their bosses just listened. I got little feedback, but I did get some questions that were related to project management. The implication was that I was rash for committing to something so grandiose. I answered each question to the best of my ability.

Japanese culture at that time was such that employees worked for the same company throughout their whole careers. They weren't in the habit of job shopping for promotions or more work satisfaction. Given that, Japanese employees are culturally hard-wired to be successful in each stage of their careers, and therefore focus diligent, practical efforts on productive achievement. So here I was, requesting that our company build a chip which my Japanese peers didn't even know could be fabricated—and *they* were the ones in the semi-conductor business! Furthermore, I was committing to a very risky schedule. In fact, almost everything I was doing was foreign to their culture, and they could have chalked it all up to the fact that I was just one of those "crazy Americans." But they still took seriously what I presented; they had to see it as feasible if they were going to allot money to my project.

Wednesday I left the rest of my group and headed home alone, knowing I'd given it my best shot and that now I needed to concentrate all my energies on the surgery before me. A head taller than most of those on board the Japanese jet, I was the only one (it seemed) whose hair wasn't black, and I was definitely the only woman on board besides attendants. I could speak to no one. I wanted to sleep; my surgeon had said I needed to arrive home with a full 24 hours of rest before the surgery and already I was cutting that short. But it was hard to sleep. I saw no reason to try to do work; without clearance for funding, there was nothing to be done. I didn't feel like reading; I couldn't concentrate. I did some yoga meditation, trying to slow my body down from the intensity of the meetings I'd concluded. In my head I knew it was time to get my body ready for this "invasion," but I was used to living primarily in a world of logic and precision, and none of that would help my internal world prepare for the surgery. The whole trip home from Japan seemed to be happening in slow motion. I arrived in time for surgery on Thursday.

Friday morning my boss called from Japan. We had been allotted the $6 million; my project was a "go!"

No other news could have helped my recovery more.

Chapter 13

Passion is energy. Feel the power that comes from focusing on what excites you.

—Oprah Winfrey

PUTTING IT ALL IN PERSPECTIVE

When I got the call that my project had secured its full budget, I thought that was the best news I could have had. But soon I was to realize that a clean bill of health for both Ken and me would be even better news.

Ken and I shared a walking regimen, enjoying the time together as we covered a three-mile path near our home. Around the time of my surgery, Ken experienced a slight tightness in his chest when we walked over hills or made major exertions. That was unusual, but we pretty much ignored it, as it went away when he stopped pushing himself. One day, all at once the usual walk didn't work for him. He turned around after a mile on flat land, and the tightness persisted even while he was only lightly stressing his heart. This was very abnormal, and couldn't be discounted. We raced to the doctor.

Initially, the doctor saw no problem but then, just to be sure, administered a stress test. When the results came in, the urgency was clear; Ken was hospitalized and a stent put into his heart. Then it was my turn to minister to him; after my post-op doctor appointments I would stop to buy books to take to the hospital to keep him occupied. Together we were grateful beneficiaries of generous help from our friends, which was what allowed me, especially, to persevere through the breast cancer surgery and then the post-op radiation. The way Ken and I recuperated from our surgeries was to go back to work.

Since I needed daily radiation doses throughout the six-week period that followed my surgery, I took the 6:30 a.m. train into Cambridge each day so that I could work during the ride in. I didn't trust myself to drive those distances. Often, I was able to work with up to six engineers who also took the train, but I only worked with them in the mornings. The evening ride home was always their own time!

I was in my office by 8:30 a.m. and would work hard until 2:30 p.m., when I left to go via "The T" and a bus to Mount Auburn Hospital in Cambridge for my treatments. After waiting for the treatment and getting my minutes with the radiation oncologist, I would walk back to the station to take the train home. I'd arrive about 6:30 p.m., exhausted. Again, friends made all this possible. Friends from our church made us dinner every other night; I'd get home to find a cooler on the front step filled with hot food ready to eat. I'd eat something and then collapse into bed. Ken was left to do the dishes and get us ready for the next day. This was a long six weeks for him.

Engineers at work felt somewhat helpless to support me, but were kind and did the best they could. They bought me a plant, not knowing that I'd killed every office plant I'd ever had—I never remembered to water them! Those engineers were ready to continue to include me in design meetings, or to back off. No matter what, though, they always kept me fully informed. They didn't know how supportive I felt they were being, just by their keeping the project on track. So while I only worked in the office six hours a day for that period, I was careful to make each minute count, scheduling continual reviews of the designs and ideas so I knew we were going in the right direction. I wondered every once in a while if the engineers might have even enjoyed my leaving them alone for more than two hours a day.

I profited from studying those Mount Auburn radiologists and radiation oncologists. As a patient, I kept asking questions—more, probably, than they ever wanted to hear. Before long I understood how much they would profit from using our new technology. Now I knew what we were building this volume-rendering system for!

By the time the six weeks were over I was back to full tilt, except that now I didn't come in on Saturdays anymore. Doctors had told me that they didn't know what causes cancer, but they could suggest this to their patients: whatever your lifestyle has been you should change it, because it was a lifestyle that had caused the cancer to grow. But I was experiencing so much enjoyment from my work that I didn't feel my lifestyle of working hard was my issue. However, staying home on Saturdays did give me extended quiet time to think about our project, and sometimes write presentations or try to understand what tasks we couldn't do. And it saved me a day commuting.

My kids were all out of college at this point. One was doing well in Colorado and the other two were pursuing their lives in New England. My husband saw me nights and weekends, but I wasn't good company: I was always thinking work. My engineers and I lived the product we were building. I flew to Japan for reviews, and also scheduled them in Cambridge with visiting Japanese colleagues. We all kept struggling to find ways to make this entire system work on a single chip; every engineer knows what that entails.

Ken's health issues were not over; by April he had two stents in his heart. I was in Japan in June when he felt his chest tightening again on his way home from work; he drove himself to the hospital. When Ken could call, he told me from his hospital bed that the doctor had put a third stent into his heart, but had declared it unsuccessful. He would need open heart surgery. Immediately. How quickly could I get back from Japan?

It was 5 a.m. on Thursday in Osaka when Ken's call had come through, and I got a Japanese colleague out of bed to help me negotiate transportation. I needed to get to Tokyo to catch a noon flight back to the States. My colleague accompanied me to the airport, explained in

Japanese the gravity of the situation and got me on the first leg of my flight back home. That plane left at 9:30 a.m. and got me to Tokyo, but landed me in the wrong airport from the one I needed for the flight back to the U.S. Speaking no Japanese, I somehow made it to a bus that took me to the correct airport. By then, my plane was so close to departure that the attendant walked me through security and all the protocol to speed up my getting on the flight. I didn't realize that they were holding the plane for me until I was on board. Walking through first class, I heard the booming voice of an old colleague from Avid who happened to be on the plane holler, "So it's *Bev* that is holding this flight up and keeping us from getting home!"

My tickets only got me to Chicago; in her haste the Tokyo attendant hadn't routed me all the way to Boston, but the stewardess was able to make a call and have a ticket waiting for me as I exited the plane in Chicago. With a short time between connections, my friend from Avid helped me stay sane until I could get onto the next plane. Meantime, an urgent case had bumped Ken's scheduled surgery back a half-day, but he was kept in the hospital. After that long stretch of international travel—Osaka to Tokyo, Tokyo to Chicago, then Chicago back to Boston—and the time changes on top of it, I arrived at Lahey Clinic in Burlington, Massachusetts, just before my husband went into surgery early that Saturday morning. Our children were fantastic, joining me there to wait for Ken's surgery to be complete.

In the weeks that followed, with his quadruple bypass behind him, Ken did well. He was recuperating quickly, but I couldn't stay home with him—I did need to be at work. Our project deadline and the difficulty of our task stayed in my mind all the time, as did my husband's condition. Friends came over to stay with Ken while I was at work. Even one high school student would stop by and keep Ken company. We were grateful for the attention. I remember thinking that, however much we want to control our lives and what we do and how we do it, we can't control it all. If we have friends and colleagues we can lean on, we are fortunate and life works out better. The sharing of situations can make them doable.

154

Sticking to our grueling schedules, our engineering team reached the day when we could let go of the design and send the chip to fabrication. We had chosen IBM for that process because they were just starting to use copper technology, then the fastest. I had experience with this fabrication plant. We were able to take advantage of their superb technology to get all our design on a single chip, which had been our goal. The fabrication was completed in about five months, with numerous treks to Vermont by our team members to clarify the design. We had boards and software ready and by May of 1999 we had put our system together and shipped the world's first single-chip, real-time 3D volume-rendering Application Specific Integrated Circuit (ASIC) Engine. Remember that I didn't join Mitsubishi and we didn't begin the formation of the Volume Graphics team until November 1997. We had given ourselves 18 months, and had made it: we had come in on time and within the budget for a product that had never been made before!

One of the reasons we made our deadline was that we were able to ship our first chip—it did not have to be refabricated! We had understood clearly that we would not make our deadline if we had needed to refabricate. Our engineers were smart, super conscientious and dedicated. We had reviewed, studied and re-reviewed our designs to optimize them and to ensure correctness. Ian was our senior designer; his ability to simplify all the complex things we were doing contributed greatly to our accomplishment. The main reason for our success was that our entire group worked as a smooth team, each person knowing exactly what to do to make everything happen by the deadline date. The system administrator was just as important as the senior designers, because the systems needed to be *up* every day. No one reneged on delivering his or her part of the puzzle. All the engineers knew what the system depended on them for. No one let us down.

RECOGNITION OF A JOB WELL DONE

At that time our team was still the Volume Graphics group of Mitsubishi Electric of America Research Labs, part of Mitsubishi Electric of America. We were proud of our technology, as was the company. Top executives at Mitsubishi Electric had allowed us to use an American fabrication company instead of having our chip fabricated in Japan even though the company did have its own semiconductor group there. I had struggled with the decision whether to fabricate the chip close to Boston, using IBM and its new copper technology, or use our semiconductor group in Osaka, which would have been cheaper for the company, of course. For a while, we made them really nervous. Our decision to use IBM was almost a "sigh of relief" for the company's semiconductor team: *they* didn't promise to complete chips on 18-month schedules! They hadn't done anything this complex. So they were happy when we decided to create the first version of a chip with another company and then to come to them to make later chips in quantity once we'd perfected what we were doing.

Every engineering project is asked to be "good, fast, and cheap." With any other project I'd worked on, the team would say "good, fast, cheap: pick two"—because it is so difficult to do more than two of those well. Good and fast costs more. Fast and cheap means you trade off quality. If you're willing to compromise on the quality, the project can finish faster at a cheaper price. But on this project we were allowed to plan and execute with our own methods. We had accomplished what we set out to do within our 18-month window and within the planned budget. Of course, six million dollars wasn't "cheap," but it was inexpensive for what we created. The real achievement was that the product worked as planned. Wow!

Each year, Mitsubishi Electric of America chose a number of products in the company to receive the "President's Award," the highest honor the company bestowed. In 1999 our Volume Graphics team's 3D chip was one of only seven projects to receive this distinction, in our case for the outstanding technology we had produced. The other six recipients received their awards for business achievements because their products had brought the company a lot of profit.

I was so proud of my team as I went to Japan to accept the award on their behalf, the first American ever to stand before Mitsubishi Electric of America's Board of Directors to receive this honor. I was so appreciative! I wished all my project colleagues could have been there to be recognized for all they'd accomplished—the team in the U.S. who had pulled this off, the research team who had done the initial work and then founded us and advised us and kept an eye on us, and those people on our team in Japan who kept everyone informed of our progress and our achievement.

Off and on during my time at this company, people would ask me why a woman would work for Mitsubishi Electric. Could I be promoted? To what position? No American had as yet been on their board of directors, but Americans were working very successfully with Mitsubishi in Japan. I never felt that I had been noticed by the powers-that-be particularly for being female. I was probably more often seen as "one of those crazy Americans," but I was always just a Mitsubishi Electric employee and part of their team and I was made to feel that I was an important part of the company. In Japan, I observed a few female engineers, but I interacted mostly with men, which is what I'd always done in America as an officer of a company. And for that matter, I hadn't seen women on boards or in senior positions very often in the United States either, so the experience with this company had felt no different. Mitsubishi valued those who delivered what they had promised, and when I did that I was taken into their corporate culture and treated like a valued employee. My gender never rose to the level of being an issue.

DIVISION OF LABOR

By the time our chip was near completion, we had hired a General Manager to maximize our revenues. He had a hard job! We had something for him to sell that worked well. But, his reasoning went, if we could build *this* chip, then a more powerful one would sell even better! Initially our team had planned with Japan that the next chip would be a modification of the first chip; I'd hoped that just refabricating it would make it strong enough to base more product on it. Instead, our General Manager felt that it made better business sense to design a new second chip that was even more capable of doing a lot of very difficult things. The GM and I argued a lot as we got going. He felt the product required much more in features, capabilities and speed to attract and generate the business that the company wanted. I preferred just modifying this chip because the fabrication would amount to minor changes and not a start from scratch. Eventually, the initial refabrication idea for the next chip got scrapped in favor of a complete redesign, in order to get the functionality that our new General Manager felt the market needed.

As often happens in business, though, we'd have to redouble our efforts: our key hardware designer, along with several software engineers, would be working with customer input to optimize the functionality of that first working chip in software, and at the same time we'd all be designing the new, second chip. I found myself applying lessons I'd learned from my years in engineering, lessons that turned out to be instrumental to our progress in working on two chips simultaneously: focus on the important things, be sure of each part of the design before moving forward, and don't be burdened with details—but don't drop them either.

My concern was how to distribute the work effectively among our very talented but very lean labor pool. Ian, our main designer on our first chip, was an engineer who built simple, elegant systems. Very few engineers can do this. It then complicates the task even more when those engineers are asked to work against their abilities at simplifying, and instead are urged to add increasingly difficult and complex requirements to the product. Besides that, it was a bit of a burnout issue as well. With the second chip we couldn't use Ian again as a lead engineer; he was needed to ensure that software modifications matched what the hardware could do, and the software brought out that chip capability. And frankly, we needed Ian's knowledge as lead designer of the first chip to help optimize that first product.

As I worked on the re-design with the other engineers—a few, very smart individuals—I saw that the only thing they were missing was this ability to simplify, simplify. We designed and built another extremely complex chip from scratch. It got more complex as we went along in order to meet all the requirements, and we were very proud of it. But the complexity cost us time and effort.

The entire scenario reminded me of a time in 1985 when I had two software engineers working on a software project. Both were very smart, but Pete was more outgoing and well known for his ability to do absolutely anything you needed doing. When he was given an assignment, he would frantically write code and within a couple of days he'd have *something* working. He'd check the code and if significant parts of it wouldn't work, he'd focus on those, and get a little farther. Then he'd add more software, firm up loose edges, test it and add some more software. He'd enhance, check, find the flows, add more software, and repeat until at the six-week deadline he had his software working as promised.

The second engineer on the project, Sally, just sat in her office for at least three weeks and "thought about it." No one saw any code in that time, but there was a lot of thinking going on. Nothing became visible. More thinking. I'd start getting worried. Sally would jot down ideas and look as though agonizing thoughts were going through her mind, but she wasn't satisfied yet to begin coding. When she finally did get to writing—probably four or five weeks into the six-week deadline—her code was simple, elegant, and worked perfectly the first time. She was the only engineer I found who didn't know how to use our debugger; she so seldom needed it.

Which engineer was better? I know that I liked to maintain Sally's code, as it was so clean and precise and took so few lines. But I valued Pete's more complex way of working, too, because he could also build things that worked. Both were fantastic engineers. But their code was the result of how each of these engineers' heads worked. I had to put them on projects that optimized the way they thought *and* the way they worked.

Back in the 1980s I spent eight years on the computer committee for my kids' school, and in the process of choosing computers for their schools over that time I got to know some of the other committee members well. One night as two of us walked to the parking lot, Stan, an engineering manager for a government project, said to me, "How many lines of code do you expect from your engineers each day?" I immediately thought of Pete and Sally. If a good amount of thinking produced 1,000 lines of code from one programmer and fifty elegant lines of code that did the same job from another, then from my manager-point-of-view I wanted the fifty lines of elegant code. I *couldn't* have rated them solely on lines of code they wrote! It wasn't even a factor for me. The things Stan felt were important weren't relevant to me. He was clearly a mostly bottom-line kind of guy; I felt it more essential to assess value from a broader perspective.

Now, in 1998 at Mitsubishi, we were being asked to build this new complex product in well under two years, and even if we had felt comfortable with that time frame, we still would have been caught by the complexity of the technology. I'd had one of those simplicity engineers on Version 1 of our Volume Renderer, yet on Version 2 the task was so complex that engineers were intensely holding on to all the pieces—unable and, in some cases, unwilling to see ways to simplify. Which didn't mean we couldn't do it. But we had to watch carefully that our engineers didn't bite off more than could be completed. As operations and requirements on the chip got more complex, the difficulty of writing code increased exponentially. In management, getting to delivery is a matter of choosing the value of various aspects of our product. How do we trade off between code quality, speed of performance, time to ship, usability, documentation, etc.? It was as if I was back in 1985 and had been asked if I wanted to work with Pete or Sally—still an untenable choice to consider.

In the end we got the job done. We weren't able to shorten the time to delivery, but we produced a powerful Version 2, which was also fantastic technology. As far as using elegant simplicity, I'm glad we used it to make Version 1. It allowed us to have a Version 2.

By this time, our Volume Graphics group had plans to spin out of Mitsubishi Electric of America, as would any startup. Mitsubishi was supporting us. Checking the numbers that Sales was predicting, I saw they were being cautious. Even though we had good customers (including a major world-wide CT scanner supplier!), we didn't have enough revenue to continue funding chips at the $6 million level. It was obvious that Version 3 would not involve a new chip, but only additional software to take advantage of the many facets of the chips built for Versions 1 and 2.

No new chip meant we'd need fewer people. When that happens, I feel the right thing to do is to cut people from the top of the organization and make it smaller. With no new hardware to redesign, my role was no longer necessary; others could manage the continuation successfully. The Volume Graphics group had challenged every aspect of my management and design engineering capabilities, but now the team was stable and could move forward from an established base.

I hadn't sent out a résumé, but I'd continued to receive some unsolicited job offers. I followed up on one, and decided to accept an offer that would shift me into a different experience. Cancer does that to you. You have to maximize the hours of every day, and do work that challenges you every day. Except for the fact that, again, I loved the people that I'd been working with, I was ready to move on.

TIP # 10: TAKE "IMPOSSIBLE" OUT OF YOUR VOCABULARY.

There is nothing you can't do. It may mean modifying your original concept, but if there is something important that needs to be done, just dive in until you figure out a solution.

Chapter 14

There's a trick to the Graceful Exit. It begins with the vision to recognize when a job, a life stage, a relationship is over – and to let go. It means leaving what's over without denying its validity or its past importance in our lives. It involves a sense of future, a belief that every exit line is an entry, that we are moving on rather than out.

—Ellen Goodman,
American journalist and
Pulitzer Prize winning
columnist

A NEW EXPERIENCE, WITH HELP FROM THE PAST

As the world moved from the 20th to the 21st century, computers continued having more and more processing power. Software was still a "craft" that needed careful attention. Hardware projects were more difficult because the technology was exponentially more complex, and the chips being developed needed to be smaller and smaller. Work opportunities in this environment were abundant. I chose to come full circle. What was to be the last job of my career was one related to the industrial control system work that had started my career when I joined Systems Research Laboratories in 1978.

It was the spring of 2000 and I hadn't sent out a résumé since leaving Number Nine, but Intellution, Inc., was courting me for the position of

Senior Vice President of Engineering at that time just purchased by Emerson Electric. The CEO/founder of the company was smart and good with business; the president was dynamic and full of energy. I looked forward to working with both of them.

I was impressed by the process control systems already in place at this company and by the engineers there who had ideas about where to carry these products into the future. I was especially pleased to find that the systems, built entirely of software, would work on *anybody's* hardware in manufacturing plants! A great concept—one I'd tried to bring to fruition at Digital Equipment Corporation when I wanted to put our VMS operating system on Intel and other platforms, but couldn't get the company's support. Here, the company *planned* it, a move that confirmed to me the perception and wisdom of the CEO. Even once Emerson Electric bought the company, Intellution could say, "Use our software on new Emerson Electric hardware, or just use it on whatever systems you have on hand." It was a powerful message. I was glad I'd be working with forward-looking people who showed a can-do attitude.

Intellution, Inc. customers relied on its software to operate without fail, because it automated and controlled the processes used to manufacture the products that customers sold. That meant our software had to be dependable, with rigorous, verified testing so that customers could use it to validate their own products. A production line couldn't have errors.

Intellution seemed to have a good handle on creating new products as well as improving their current ones. The company included a group in Germany led by Owen, a dynamic inventor who had great ideas for where we could take our software—groundbreaking possibilities, in fact. He obviously knew our market, and knew some of his ideas could position us to take over the market in the future. But he had an understated way of presenting those concepts. Even then, Owen didn't expound on his thoughts, but had to be coaxed to reveal them. I had to listen carefully when he talked about them. I learned when and how to dig out his inventive ideas when I supported him in meetings with Marketing and Sales, or senior management.

Another great colleague was our Chief Technical Officer, who was focused at the time on product directions as the acting head of research when I took over Engineering. He was also worth listening to! He had us enhancing our core products and simultaneously researching fantastic ideas to keep our customers competitive and to keep us profitable.

I felt right at home with this group. They craved management attention for their work, and they were doing interesting things. I had to understand the hardware they were working on, but make sure the software remained hardware-independent. I had replaced someone who'd been a strong engineer and a dynamic Senior VP of Engineering, and I appreciated that the company let me just step in and continue his hands-on approach. The foundation work had been done. What was interesting to contemplate was where we could now take the technology.

For me, though, technological advancement formed only part of the picture. I had always focused on economy of time and process as well, and I loved the efficient way we were able to build software during my tenure at Intellution. I would draw on all my experience from my early days up to the present, remembering how my teams and I could build anything. For several years at Intellution we engineers only got better in making our products and in the efficiency with which we produced them.

WRITING ON THE WALL

I attempted to make the most of opportunities to enhance and expand our line when those chances arose. At first I tried to buy other products to complete our offering, and then, in fact, to buy a small software company in the Midwest whose products meshed with ours. Emerson

Electric wasn't willing to let me do that. They were unable to see that little software company as a valued part of our future. I realized I was back to working with an established "hardware company," one exhibiting the same tunnel vision that I'd seen plague some of its industry predecessors. This time I didn't fight it; I just worked "deals" that let us sell this small Midwestern company's products together with our own, without buying the whole company. (Several years later Intellution, Inc. was sold, and the company that bought it also bought the little company whose products we had been using. It was satisfying to have our earlier judgment validated.)

By fall of 2000, Intellution lost our president to another company; our founder and CEO then became acting president. I had fun with this for a while, but all too soon the CEO also left. I was saddened to discover the reasons for the departures. This big hardware company had increasingly flexed its muscles in determining what the founder and president could or could not do, so when these two entrepreneurs didn't feel like they were working anymore for the company they had grown, they were out of there! This disappointed me—I'd been influenced to join the company because I really respected both men.

Their departure signaled to me that the writing was on the wall about getting to do new and interesting development. Up until that point, I'd been growing the engineering group. We had been looking at exciting research and possibilities for expanding our products to further automate the processes of manufacturing. We wanted to help design our customers' products as well as improve how we could "cookie-cutter" their final assembly. We'd implemented some changes in our products, and were filled with new ideas.

To lead us, our parent hardware company, Emerson, brought in a president of their own, a man who had headed a marketing group at Westinghouse, another Emerson company. He was being brought in to drive Intellution, but from the beginning, his goal seemed to be selling off our division. For the most part, the engineering group stayed in its own area, and I was able to continue to drive new products and improvements to existing products, which I loved doing. Granted, that became a bit more difficult because our budgets were being reduced, a situation that always restricts—or at least reduces—the flow of ideas. So I wasn't getting to bounce ideas off the executive officers. But I stubbornly kept focusing on finding innovative solutions: working with new problems and ideas was what had always inspired and sustained me. And it was keeping us profitable.

For a while, the new president/COO seemed content to pretty much let me run Engineering with no input from him, as our group was doing so well for the company. But then he began coming to me with comments and his own ideas of how to keep this company efficient.

For example, he was appalled that he would come to the plant on some Sundays and there would be few, if any, engineers at the office. I explained that our engineers worked hard and they worked smart, which didn't necessarily mean they had to be in the office more than 60 hours every week. When a product ship date was imminent, all engineers worked nights and weekends; I'd often buy pizza for those still on the job at 7 p.m. But when we made our deadlines, I was happy to see engineers enjoy a balanced life, going off to play in their band or attend events, or just see their kids and their spouses. But the new COO didn't see it that way.

When he confronted me about it, I just stood up to him. Funny, I guess I'd had enough years behind me. I told him if the work wasn't getting done, or if the products were faulty and failing, then he should hold me responsible. Meantime, he had to let me manage my engineers as I saw fit. I'd never rated engineers by "hours worked," and I didn't intend to start.

It reminded me of how I felt in 1984 getting the ludicrous question, "How many lines of code do you require of each engineer each day?" I felt the same when my COO basically asked, "How many hours do you expect an engineer to work each week?" If my engineers could build simple, elegant code in five hours a week, I'd prefer that to them working very hard writing lots of code throughout a 75-hour week. Work smart, not hard. In my experience, if I depended on my engineers to work smart, and they knew my expectations, that's what they did.

I began sensing a subtle shift in my junior engineers around this time. Since 1978, when I'd started working in engineering, any engineer could quit at any time and he could immediately find another job, usually at a higher salary. But in 2000 the world economy was stalled, manufacturing companies in the U.S. weren't making money, a lot of manufacturing was being shifted overseas, and engineers were worried about keeping their jobs. Unspoken doubt began to hover in people's minds at Intellution: What if this COO didn't keep us solvent and engineers would have to be laid off? This worry manifested itself in the workplace, as engineers became hesitant to take the risks that previously had been second nature to them in building strong products.

The uncertainty was insidious and damaging. An engineer who worried about his job moved too cautiously, and I watched the young ones in particular do just that. I knew our company wasn't going to move fast enough to continue leading the market if engineers didn't feel they could try things that might fail. And the shaky economy was promoting this behavior. The United States had always been a place where engineers could "go for the gold" and not have to worry about achieving success at everything they did—they'd be good enough often enough that their company would know their worth, and overall their results would be awesome.

I had never employed engineers who didn't just build products for the experience of building the best thing out there. But now I was managing these guarded, more tentative engineers. I struggled to make them feel "safe."

After several years of fun at Intellution, I saw that the company was about to be sold again. Our president had just asked me to cut another $600,000 from the Engineering budget, and my group was not the only one. From his perspective, the requests seemed rational: the more the budgets were reduced, the better we would look as a profitable enterprise, and the more salable we would be. During the less than two years I'd led Engineering I had been asked to downsize the group several times, so I'd already frequently reduced my headcount. To make this new, deep cut, though, I knew I'd be forced to give up key personnel—I had nothing left to cut in my budget except salaries.

The thought came to me: What if I started with my own salary? That would go a long way to save the jobs of engineers who were critical to daily development. I calculated that, when given this extremely pared down budget, one of the managers under me could handle our team. We were already smaller: I was managing fewer people than I'd been hired to manage, and we hadn't bought the company that would have grown my group, so by this time we were a "lean and mean" engineering team. Even with my leaving, I would have to significantly downsize the group further to cut the full $600,000 from the budget. The remaining engineers could be controlled by a less experienced manager.

While I was weighing this line of thought, I received some unexpected support for it. My Chief Technical Officer came to me, unsolicited, to suggest that I eliminate *his* position. He wasn't sure that his role would remain interesting if he had to operate with a much smaller group. Besides, he was ready to strike out on his own as an entrepreneur and work to start a new company.

This suggestion helped me start to assess and balance which job cuts made sense. If my CTO and I remained with the company, I would have to eliminate the positions of at least six other engineers. Project management would go away, or a chunk of our testing program would be eliminated, or customer support would be unduly affected, or engineering itself would be reduced. Taking even one or two people from each of these areas would impact deliverables and would keep our group from doing new, interesting work.

On the other hand, there might be some personal advantage in staying on. I had already been given favorable feedback on the smooth workings of my engineering group from the CEO of a company that was trying to buy us. After reviewing all of Intellution, Inc., this CEO had spent several hours with me and had shared that, if he could, he would buy only the engineering part of the company. I was fairly confident that no matter who bought Intellution, I would be given expanded opportunities under new company leadership.

I pondered.

One of the things I'd been told when I was struggling with breast cancer was that my lifestyle was obviously one in which cancer had been able to thrive. "Change your lifestyle," they said, "whatever it is." So, I had quit working on Saturdays, except when really needed. That was my only concession at that time to a changed lifestyle.

It had been four years since that experience. Maybe, I thought, it was time to really change. To retire and keep my time as my own. To spend time with Ken; my job had limited my ability to travel as we wanted. I didn't have time to write or see grandchildren, to pursue other passions or to do many things that are outside of "work." And, I reasoned, if my mind needed the satisfaction of some high-tech interaction, I could still consult on occasion. Yes, I began to feel the time had come to leave my world of technology, to step out of the whirlwind that was hardware and software development.

In 2002, I decided to do just that. I listed my position among those to be eliminated.

But nothing is simple and straightforward in getting into, staying in or getting out of business. The president didn't accept my name on the list—not only because Engineering was running well and he didn't want to spend time on that, but because evidently I was part of the attractiveness of the package he'd counted on for company sale opportunities! He pushed back, but by then I'd made up my mind. In the end we compromised on timing: I stayed a while longer to train a replacement, but I was numbering my own days at Intellution.

Had I cut myself out at the expense of my own career? Maybe. But I couldn't imagine a *better* career than the one I'd already experienced. Thankful for that career, I was ready to turn the page, content with leaving good engineers to carry on.

Epilogue

...The credit belongs to the man who is actually in the arena, whose face is marred by dust and sweat and blood; who strives valiantly; who errs, who comes short again and again, because there is no effort without error and shortcoming; but who does actually strive to do the deeds; who knows great enthusiasms, the great devotions; who spends himself in a worthy cause; who at the best knows in the end the triumph of high achievement, and who at the worst, if he fails, at least fails while daring greatly, so that his place shall never be with those cold and timid souls who neither know victory nor defeat.

—Teddy Roosevelt
26th President of the
United States

I never had a job where I felt I could just coast—working with leading-edge technology meant taking high risks and living within ever-changing environments. My most rewarding jobs brought the hardest challenges. Every aspect of my career involved change.

I was always moving out of my comfort zone to build something I didn't know how to build, create something I'd never heard of, or invent something that didn't exist. But when I was doing anything with a group of smart people, I never I thought we *couldn't* do it—mostly, I just didn't know *how*. My job was always to get input from a lot of people and then work with the ones who had the strongest ideas to define and mold what we would do and how we'd go about it. But even when engineers do a tremendous job, companies change, the world changes, the next opportunity changes. As we went along we changed our path or process if that was necessary to build the best products in the best ways.

Often I felt that software and hardware engineering management was like acting in a play. No matter how good a role is, the next one will be different; it will afford new challenges. You have to ask yourself: Do I want this part? If the answer is yes and you're offered the part, you give it your all until the curtain goes down. Then you find another challenge. And start again.

In the late 1980s, the head of software for Digital Equipment Corporation, who managed more than 6,000 engineers, included me on a committee that was studying "change." He reacted to a presentation I'd given about the important roles of "pioneers" and "settlers" in technology, and how the software industry desperately needed to value both kinds of engineers. "Pioneers" were the ones who would trek out into the unknown and, pushing technology forward, find new places to go with it. "Settlers" would maintain support, fix bugs and enhance the technology in existing products, using technology to make a more comfortable existence for everyone. Before that boss left our company, he said to me, "Bev, you and I were wrong about pioneers and settlers. Technology is moving too fast—we need to have *all* pioneers!"

He was right. And I certainly tried to do my part.

I was fortunate in my career to have been able to pioneer new technology. The rapid pace of innovation in the computer field motivated and energized me.

But I was also fortunate to be a something of a pioneer myself.

I was among the early people in the industry who managed the combination of hardware and software technologies to create better, simpler, more useful products. And as the value of the contributions of software engineers increased, I was one of the first women to manage those engineers and rise to an executive level in that management.

Maybe it was a matter of being in the right place at the right time, but I brought my blend of boldness, talent, discipline, care, respect, openness and good humor—my way of engineering out loud—to the rising wave of the computer revolution in the last quarter of the 20th century.

I am grateful to have had the opportunity to ride that wave and glad to have made the journey.

TIPS FOR GOOD SOFTWARE AND HARDWARE ENGINEERS

TIP #1: DON'T SETTLE FOR LESS.
If things don't turn out as you planned, do what you need to do in the job you have, but keep your mind open for possibilities. Dwell on those as you continually seek the best job-fit for you.

TIP #2: KEEP IT STEADY.
Take one task at a time and do it well. Or break the task into pieces until you can make it work. As you solve the problem before you, the next hurdle becomes easier. You're never in "over your head." You may just need to learn to hold your breath until you can surface.

TIP #3: PLAY WELL WITH OTHERS.
As a manager, you cannot do everything yourself. Understand and utilize those around you; persuade them to take the actions you need them to take. Convince them that an idea is theirs! Engineers are a quiet bunch and fellow managers are worlds unto themselves. Learn to design with the former and collaborate with the latter.

TIP #4: ENJOY THE GOOD STUFF.
Tell your body that this is the work you're built for, and revel in what you're doing. Help the body thrive, and absorb an attitude that says, "This is good. I like what I'm doing."

TIP #5: BALANCE LIFE.
You'll do your job better if you can enjoy other creative outlets and let your work simmer in the back of your mind, getting inspiration from other things you do.

TIP #6: BE YOUR OWN CHEERLEADER; TOOT YOUR OWN HORN.
Sometimes fantastic work goes unnoticed. Push your products, ideas, thoughts and inspirations. Sometimes you can push quietly, through

your code or designs, but other times you need to do whatever it takes to be noticed! If company "fires" get all the attention and you have a group that runs like a well-oiled machine, you have to publicize yourselves.

TIP #7: CHANGE WITH CHANGE.
Change happens. Go with it, and don't let it get you uptight. You can't keep things the same, and you don't really want to. New challenges are workable.

TIP #8: RESPECT YOUR MANAGER.
No matter how great the job, you must respect your own manager in how s/he thinks and works. Respect what your manager values. If you cannot, you need to find yourself another manager.

TIP #9: STOKE YOUR FIRE.
You get stale if you stay in the same job for too long. Whether you take a series of jobs with different companies or move within a company, try to make each new challenge stretch you. Find ways to make yourself think outside the box.

TIP # 10: TAKE "IMPOSSIBLE" OUT OF YOUR VOCABULARY.
There is nothing you can't do. It may mean modifying your original concept, but if there is something important that needs to be done, just dive in until you figure out a solution.

TIP # 11: MAKE THESE TIPS YOUR HABITS!

www.ingramcontent.com/pod-product-compliance
Lightning Source LLC
LaVergne TN
LVHW092005050326
832904LV00018B/324/J